Sebastian Seiffert
Physical Chemistry of Polymers

Also of Interest

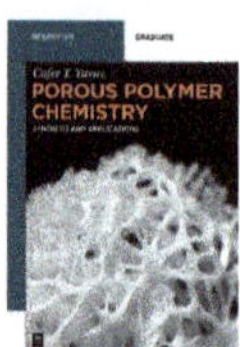

Porous Polymer Chemistry.
Synthesis and Applications
Yavuz, 2021
ISBN 978-3-11-049465-5, e-ISBN 978-3-11-049468-6

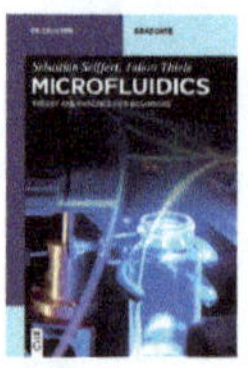

Microfluidics.
Theory and Practice for Beginners
Seiffert, Thiele, 2019
ISBN 978-3-11-048777-0, e-ISBN 978-3-11-048770-1

Organic Chemistry.
Fundamentals and Concepts
McIntosh, 2018
ISBN 978-3-11-056512-6, e-ISBN 978-3-11-056514-0

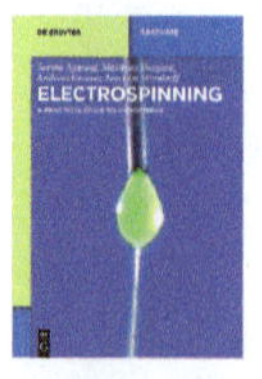

Electrospinning.
A Practical Guide to Nanofibers
Agarwal, Burgard, Greiner, Wendorff, 2016
ISBN 978-3-11-033180-6, e-ISBN 978-3-11-033351-0

Polymer Surface Characterization.
Sabbatini (Ed.), 2014
ISBN 978-3-11-027508-7, e-ISBN 978-3-11-028811-7

Sebastian Seiffert

Physical Chemistry of Polymers

A Conceptual Introduction

DE GRUYTER

Author
Prof. Dr. Sebastian Seiffert
Johannes Gutenberg University Mainz
Department of Chemistry
Duesbergweg 10–14
D-55128 Mainz
Germany
sebastian.seiffert@uni-mainz.de

ISBN 978-3-11-067280-0
e-ISBN (PDF) 978-3-11-067281-7
e-ISBN (EPUB) 978-3-11-067284-8

Library of Congress Control Number: 2019954092

Bibliographic information published by the Deutsche Nationalbibliothek
The Deutsche Nationalbibliothek lists this publication in the Deutsche Nationalbibliografie; detailed bibliographic data are available on the Internet at http://dnb.dnb.de.

© 2020 Walter de Gruyter GmbH, Berlin/Boston
Cover image: Sebastian Seiffert
Typesetting: Integra Software Services Pvt. Ltd.
Printing and binding: CPI books GmbH, Leck

www.degruyter.com

Foreword

Polymer science is a field that requires good knowledge about both chemistry and physics. This is because it is chemistry that sets the specificity of each polymer, but behind that, it is physics that sets the universality of the properties of *all* polymers. The convergence of both lies the ground for a huge variety of applications of polymers. In such, though, a user (=a customer on the market) usually doesn't care about the material itself, but instead, only about the *function* that it provides. Thus, material designers (be it in industry or academia) are demanded to develop materials that provide functions of interest, whereby customers, however, do not actually care about the beauty of the material that does that, but instead, only about its utility. It is therefore obligatory for material designers to translate the desired functions (=the customers' wishes) into measurable physical material parameters, and then, to further translate these parameters into chemical structures. This translation can only be done on the basis of a sound understanding of structure–property relations of polymers. And that is what this book is targeted at. Physical chemistry of polymers intends to bridge the structure of polymers, as provided by chemistry, to their properties, as captured by physics. Once this bridge is erected, it can be passed in either direction, thereby allowing us to understand from what structural characteristics a certain beneficial property comes from, or vice versa, to predict what structure will provide a property (and therewith a function) of interest.

A challenge in this endeavor is that at a first view, the concepts and treatments in physical chemistry of polymers appear to be rather new and unknown to students. It is therefore another goal of this book to familiarize its readers with these approaches, and to demonstrate that they are actually not at all new, but instead, quite related to concepts known from elementary physical chemistry. Hence, in this book, many relations are drawn to such classical physical chemistry contents, and the whole focus of this book is generally more on conceptual universality rather than on detailed specificity. For the purpose of the latter, all illustrations in this book are styled as lecture-like schematics. The content is ordered into six chapters that arc from fundamental polymer physical-chemical principles to actual structure–property relations, including a touch on experimental approaches to characterize both of the latter. On top of this six-chapter structuring lies a secondary structure that portions the content into 18 lesson units, which could be conceived as 90-min lectures each. The author teaches those units in two sequential classes, Physical Chemistry of Polymers 1 on bachelor level (Lessons 1–9) and Physical Chemistry of Polymers 2 on master level (Lessons 10–18). With a semester length of commonly 12 weeks, this portioning gives ample time to not only cover the respective content, but also to leave some flexibility for additional lessons to be filled with in-depthment, questions and answers, and a practicing exam. Alternatively, in a long semester of 14 weeks (as it is the case in the winter term in Germany), the full content can be covered if four lessons are left out; for example,

https://doi.org/10.1515/9783110672817-202

these may be Lessons 9, 12, 17, and 18 if there shall be no such big emphasis on analytical characterization of polymer systems, or these may be Lessons 12, 15, and 16 (plus one more) if there shall be no such a deep focus on rheology.

The year of appearance of the first edition of this book is the inspiring Staudinger year 2020, the year of polymers. Based on this big anniversary of our field, this book aims to converge chemistry and physics of polymers and to make it understandable and appetizing to students at levels as early as possible.

Mainz, spring 2020

Contents

Lessons

Literature Basis

Isaac Newton once pointed out his work to be outstanding as he could "stand on the shoulders of giants" (Letter from Sir Isaac Newton to Robert Hooke; Historical Society of Pennsylvania). In the same sense, this textbook is based on the following seminal existing ones:

M. Rubinstein, R. H. Colby: *Polymer Physics*, Oxford University Press, New York **2003**
H. G. Elias: *Makromoleküle*, Wiley VCH, Weinheim **1999** and **2001**
B. Tieke: *Makromolekulare Chemie*, Wiley VCH, Weinheim **1997**
B. Vollmert: *Grundriss der Makromolekularen Chemie*, Springer, Berlin Heidelberg **1962**
J. S. Higgins, H. C. Benoît: *Polymers and Neutron Scattering*, Clarendon Press, Oxford, **1994**

Among those, the book by Rubinstein and Colby has given particular inspiration for the present one. In the author's view, these colleagues' seminal book has a physics-based approach (as its title also says) targeted at grad students (as its backcover-text says), which would correspond to readers at a PhD-student level in the European system. As a complement to that, the present textbook provides a physical-chemistry-centered viewpoint addressing both grad and undergrad students, which corresponds to students also on bachelor and master levels in the European system.

This book is based on a script to the author's lecture series "Physical Chemistry of Polymers" at JGU Mainz. The writing of this script has been assisted by Dr. Willi Schmolke in the years 2018 and 2019, who was a PhD student in the author's lab at Mainz then; special thanks go to Willi for this assistance with the book's fundament, along with further thanks to PD Dr. Wolfgang Schärtl for proofreading it. An even deeper fundament has been laid by another person named "Willi": the author's own respected teacher in the field of polymer science at TU Clausthal in the years 2003–2008, Prof. Dr. Wilhelm Oppermann.

https://doi.org/10.1515/9783110672817-205

1 Introduction to polymer physical chemistry

Lesson 1: Introduction

As a starting point of our endeavor, the first lesson of this book will introduce you to the fundamental terms and basics in the field of polymer science, along with its historical development. You will learn what a polymer actually is, why these large molecules are interesting, and what the fundamental differences are in their treatment compared to classical materials that are built up of small molecules.

1.1 Targets of this book

The prime goal of polymer physical chemistry is to understand the relations between *structure* and *properties* of polymers. Once such understanding is achieved, *molecular parameters* can be rationally connected to *macroscopic behavior* of polymer-based matter. This connection bridges the fields of *polymer chemistry* and *polymer engineering*, as shown in Figure 1, thereby allowing polymer materials to be tailored by rational design. Even though these fields may first seem far apart and separated (by an impassable water body in Figure 1), the discipline of polymer physical chemistry arcs and connects them. (And in the end, by looking underneath the water, we realize that they are actually connected inherently.)

To achieve this goal, along our way, we need to learn some fundamental approaches in the physical chemistry of soft condensed matter; we will also learn why polymers are a prime example of such matter. First, we need to consider the multibody nature of each polymer chain as well as the myriad of different local microscopic conformations that it can adopt. This requires us to limit our focus on *averages* over all these individual states, which we obtain by suitable *statistical treatment* as well as *mean-field modeling*. Second, as a direct result of such consideration of the conformational statistics of polymer chains, we will see that pretty much of all their properties are coupled to each other in the form of **scaling laws**. Hence, scaling discussions will be a frequent means in this book. With this treatment, we will recognize that polymers exhibit both **specificity** and **universality**. As an example, the relation between the size and the mass of a polymer chain follows a power-law proportionality, independent of what polymer exactly we look at; this is universality set by physics. The proportionality factor, however, depends on the specific polymer at hand; this is specificity set by chemistry. As a further example, note that all polymers exhibit a glass transition temperature, T_g, at which they change from a hard, glassy state to a soft, leather-like state. They do so because of *universal physics* shared by all polymers. However, the exact value of each polymer's T_g is determined by each polymer's *specific chain chemistry*.

https://doi.org/10.1515/9783110672817-001

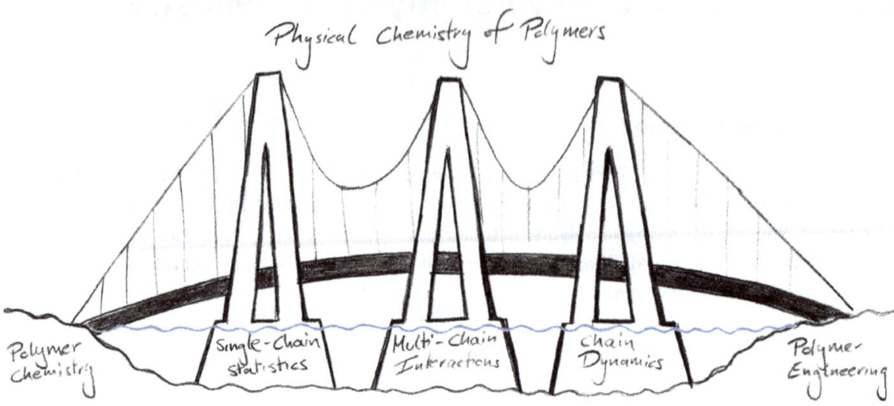

Figure 1: *Polymer physical chemistry* builds a bridge between *polymer chemistry*, which focuses on the primary molecular-scale structure of polymer chains, and *polymer engineering*, which focuses on the macroscopic properties of polymer materials. This connection allows for a rational design of polymer-based matter to obtain tailored materials. To erect the bridge, three pillars are required: single chain statistics, multichain interactions, and chain dynamics. These pillars are in the focus of the second, third, and fourth chapters of this book. The final bridge on top of this basis is in the focus of the fifth chapter, which sheds a particular focus on polymers' most important kind of properties: their mechanical ones.

1.2 Definition of terms

The term *macromolecule* is derived from the Greek *macro*, meaning "large", and the word *molecule* in its usual meaning.[1] Hence, this term refers to large and heavy molecules, with sizes, *r*, in the range of some ten to hundred nanometers, and molar masses, *M*, in the range of 10^4–10^7 g·mol^{-1}, as shown in Figure 2. Delimited from that, the term *polymer*, which is derived from the Greek *poly* and *meros* meaning "many" and "parts", refers to chain molecules constituted of a sequence of many repeating units. These units, named *monomers*, are usually covalently jointed.[2] Therefore, whereas the term *macromolecule* generally refers to any sort of large molecules, the term *polymer* specifically refers to those built of a repeating sequence of monomer units. Hence, **each polymer is a macromolecule, but not each macromolecule is a polymer**.

1 Note: the term *molecule* actually derives from the Latin *molecula*, meaning "small mass"; as such, the word *macromolecule* would mean "large small mass", which is sort of an oxymoron.

2 A rather new class of self-assembled matter is *supramolecular* polymers; it is based on chains that are *noncovalently* jointed.

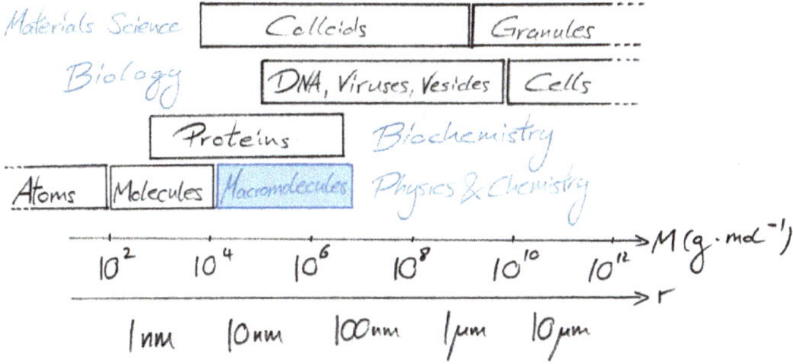

Figure 2: Molar mass, *M*, and size range, *r*, of various atomic and molecular structures.

Somewhat doubtful in this context are biopolymers such as proteins and DNA. On the one hand, one may argue that they are composed of a repeating sequence of units, which are peptides in the case of proteins and nucleotides in the case of DNA, justifying them to be termed polymers. On the other hand, despite this repetition of peptides or nucleotides, it is not a monotonic sequence of just one out of the 20 natural amino acids or just one out of the four different nucleobases that are available in nature, but a diverse variety of sequences of them along the chain. This argument may then question the adequateness of terming these macromolecule polymers. This is, however, a pea-picking discussion. Following this line of thought, even random copolymers, which are actually well accepted to be termed polymers, should not be named such. We therefore refrain from this argument. Instead, we point out that by virtue of their ability to store information in their primary monomer sequence, which in turn is a basis for their highly specific function and functionality, proteins and DNA are nothing less than the link between matter and life, that is, between chemistry and biology.

The International Union of Pure and Applied Chemistry (IUPAC) defines a polymeric molecule as follows:

> A molecule of high relative molecular mass, the structure of which essentially comprises the multiple repetition of units derived, actually or conceptually, from molecules of low relative molecular mass. In many cases, [...], the addition or removal of one or a few of the units has a negligible effect on the molecular properties.

This definition does not only apply to molecular properties but also to macroscopic properties, as can be recognized when considering the example of simple hydrocarbons in Figure 3. Starting from methane, a family of alkanes and polyethylenes is obtained by adding $–CH_2–$ repeating units. For the first 17 additions, the boiling point, T_{bp}, is raised sharply, until then, further addition leads to solid compounds that decompose instead of boiling at high temperatures. An upward trend is also observed for the melting point, T_{mp}, up to even the first about hundred additions. Then, however, the picture changes. Further addition does no longer change the

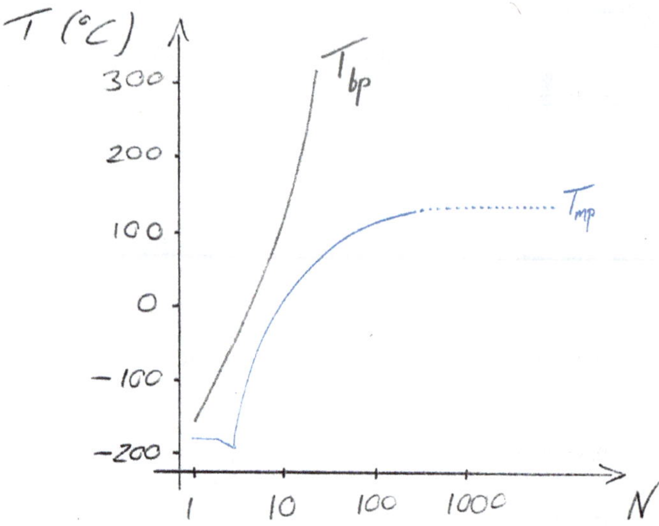

Figure 3: Boiling and melting points, T_{bp} and T_{mp}, of alkanes and polyethylenes with N methylene ($-CH_2-$) units. Whereas the boiling point increases with N and eventually becomes inexistent from $N = 17$ on (because beyond that, such high temperatures would be necessary for boiling that the molecules would break rather than boil), the melting point first increases with N but then eventually levels off in a plateau. In this plateau, addition or removal of a few $-CH_2-$ units does not change T_M considerably. Picture redrawn from H. G. Elias: *Makromoleküle, Bd. 1 – Chemische Struktur und Synthesen* (6. Ed.), Wiley VCH, **1999**.

melting point; instead, it has reached a plateau, which denotes the polymeric regime. In this regime, addition or removal of single or few $-CH_2-$ repeating units has practically no impact on the material properties such as T_{mp}. The materials' appearance in this range of high N is that of hard solids rather than that of oily or even volatile liquids at low N.

From the above notion, we may conclude that polymers are macromolecules composed of a repeating sequence of units; these units are sometimes termed *monomer units* and sometimes termed *structural units*. Note the difference between these terms: the term *monomer unit* usually refers to the chemical species from which the polymer has been made, or more precisely, to the chemical sequence along the main chain into what the actual monomer has turned into upon polymerization, whereas the term *structural unit* refers to the smallest possible repetitive entity along the chain. Often this is the same, but sometimes it is not. For example, consider polyethylene again. In this polymer, the monomer is ethylene, $H_2C{=}CH_2$, which turns into a monomer unit of $-H_2C-CH_2-$ upon polymerization, whereas the smallest possible repetitive unit, that is, the structural unit, is just the methylene group, $-CH_2-$. Similarly, in polyamide, $[-HN-(CH_2)_x-NH-CO-(CH_2)_y-CO-]_n$, the structural unit is $-HN-(CH_2)_x-NH-CO-(CH_2)_y-CO-$, whereas the monomer units are $-HN-(CH_2)_x-NH-$ and $-CO-(CH_2)_y-CO-$, resulting from the respective diamines and diacids upon polycondensation.

1.3 Irregularity of polymers

Polymers aren't regular. This is because they are made by statistical processes during most polymerization reactions, leading to inherent nonuniformity even if the basic chain-propagation reaction step proceeds perfectly. On top of that, if there is a nonperfect course of that step, each "side product" formed during it will be incorporated into the chains and create irregular local constitutions. As a result, polymers display different types of nonuniformity and irregularity.

1.3.1 Nonuniformity of monomer connection in a chain

Even along each single chain, there is structural irregularity if multiple ways of monomer–monomer interconnection are possible. One type of such irregularity is the three different possibilities of head-to-head versus head-to-tail versus tail-to-tail interconnection of asymmetric monomers such as the broad class of vinyl compounds, as shown in Figure 4.

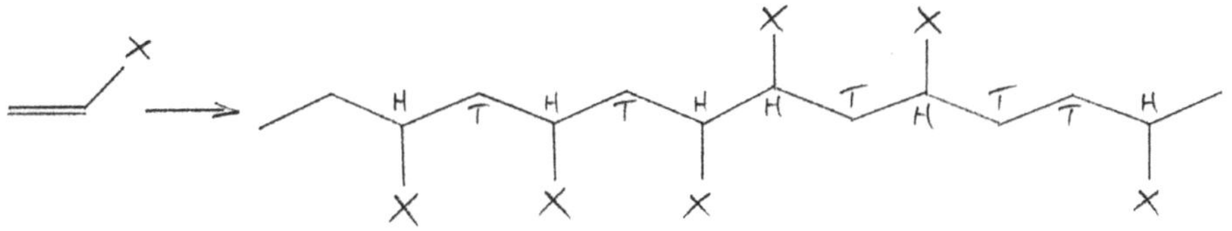

Figure 4: Head-to-head versus head-to-tail versus tail-to-tail interconnection of asymmetric monomers such as vinyl compounds.

A second type of irregularity resulting from the connection of such asymmetric monomers is different stereochemistries along the chain, referred to as **tacticity**, as shown in Figure 5. This property is of high relevance when polymers are crystallized, which does only work with isotactic or syndiotactic chains. This is because crystallization requires atoms or molecules to be able to arrange themselves regularly on a lattice. In polymers, though, this is tough, because the molecules (=the structural units) are connected to one another in the form of chains. Only if this connection is very regular, the building blocks can form a regular lattice.

In the case of diene-monomers, even more ways of interconnectability exist, as shown in Figure 6.

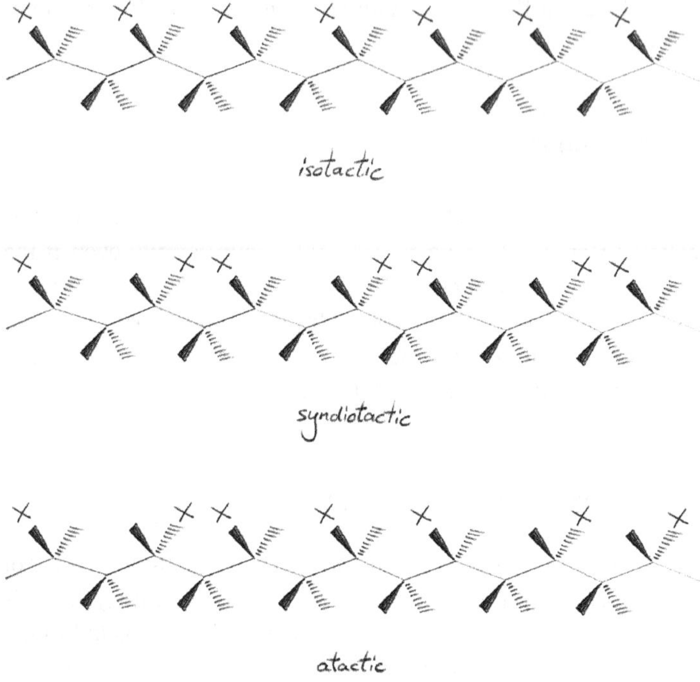

Figure 5: Different types of tacticities in vinyl polymers.

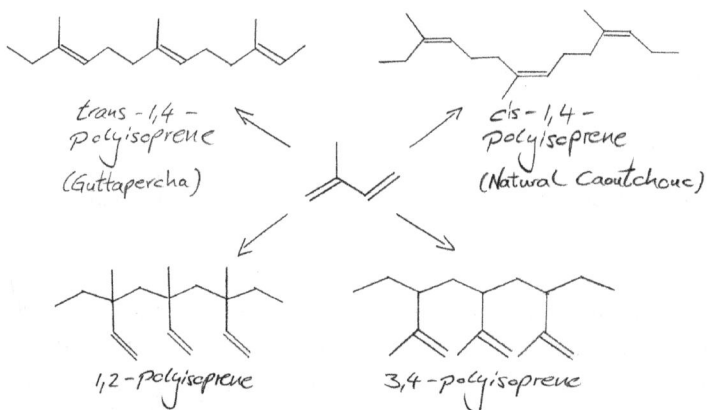

Figure 6: Different possibilities of diene-monomer interconnection in the example of isoprene.

1.3.2 Polydispersity in an ensemble of chains

On top of the local irregularity of the monomer connectivity along each individual chain, another relevant irregularity in polymer systems is chain length and molecular weight (unit: Da) or molar mass (unit: g mol^{-1}) **polydispersity** in ensembles of multiple chains. As a result of the statistical nature of most polymerization processes, the distribution of these quantities can be quite marked. A typical shape of a distribution curve, in two common ways of representation, is shown in Figure 7.

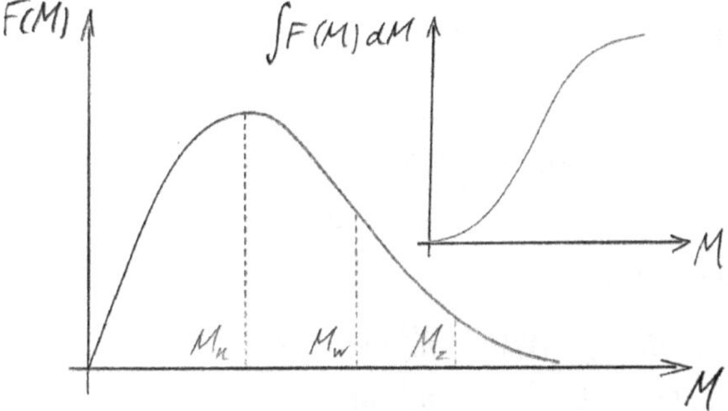

Figure 7: Schematic of a typical shape of a molar mass (unit: g mol^{-1}) or molecular weight (unit: Da) distribution in a polymer sample, represented as the frequency of occurrence (typically a relative mass percentage), F, plotted on the ordinate, of different molar masses in the sample, M, plotted on the abscissa. The upper-right inset shows a different representation of this distribution, which is the integral of all molar masses, plotted on the ordinate, up to a specific one considered on the abscissa.

For practical purposes, it is more convenient and often sufficient to not deal with the full distribution but just with some characteristic averages of it. These are the **number-average molar mass**, M_n, the **weight-average molar mass**, M_w, and the **centrifugal-average molar mass**, M_z, as represented by dashed lines in Figure 7. These averages may be calculated from the so-called moments of the distribution function, which are generally defined as follows:

Order ↘ Feature ↓

$$\mu_g^{(K)}(M) = \frac{\sum_i g_i M_i^{K}}{\sum_i g_i M_i^{0}}$$

↑ Argument; here: $g = n, w,$ or z

In statistics, the moments of a distribution quantify its characteristic properties. The first moment generally corresponds to the arithmetic average, the second moment to the width of the distribution, and the third moment to its skewness. In the formula above, the argument g denotes whether number fractions, weight fractions, or z-fractions are used to weigh the contributions of different M_i in the right-hand side calculation.

With this definition, we get the three characteristic average molar masses as follows:

– Number average: $M_n = \frac{\mu_n^{(1)}(M)}{\mu_n^{(0)}(M)} = \frac{\sum_i N_i M_i}{\sum_i N_i} = \sum_i x_i M_i$

where x_i is the mole fraction of species i.

– Weight average: $M_w = \frac{\mu_w^{(1)}(M)}{\mu_w^{(0)}(M)} = \frac{\sum_i W_i M_i}{\sum_i W_i} = \frac{\sum_i N_i M_i^2}{\sum_i N_i M_i} = \frac{\mu_n^{(2)}(M)}{\mu_n^{(1)}(M)} = \sum_i w_i M_i$

where w_i is the weight fraction of species i.

– z-average: $M_z = \frac{\mu_z^{(1)}(M)}{\mu_z^{(0)}(M)} = \frac{\sum_i Z_i M_i}{\sum_i Z_i} = \frac{\sum_i W_i M_i^2}{\sum_i W_i M_i} = \frac{\mu_w^{(2)}(M)}{\mu_w^{(1)}(M)} = \frac{\sum_i N_i M_i^3}{\sum_i N_i M_i^2} = \frac{\mu_n^{(3)}(M)}{\mu_n^{(2)}(M)}$

The number average puts the same emphasis on short and long chains in the sample and simply averages over their weight or molar mass by accounting for their number fraction in the sample. An experimental means to determine this average value is by methods that probe phenomena that specifically depend on the number of molecules in a sample; a prime example of such methods is those that probe colligative properties such as the osmotic pressure of a polymer solution. In contrast to that, the weight average puts more emphasis on the long and therefore heavy chains in the sample, as it averages over them not by accounting for their number fraction in the sample but for their contribution to the total sample weight. As a result, the heavy chains in a sample receive a greater pronunciation in the averaging. The most typical experimental means to probe this average value is by static light scattering. The z-average pronounces the heavy chains in the sample even more severely. The standard technique to probe this average value is by ultracentrifugal analysis.

We may illustrate the difference of the three types of averages of the molar mass in an example. Consider a sample of 100 chains, 10 × 100,000 g·mol^{-1}, 50 × 200,000 g·mol^{-1}, 30 × 500,000 g·mol^{-1}, and 10 × 1,000,000 g·mol^{-1}. According to the above formulae, M_n = 360,000 g·mol^{-1}, M_w = 545,000 g·mol^{-1}, and M_z = 722,000 g·mol^{-1}. Note that quite different averages are obtained even though they are all related to the absolute same sample. This is because M_n balances light and heavy chains equally, whereas M_w (and M_z even more) puts more emphasis on the heavy chains. Based on what we have said above, we may illustrate the values of M_n and M_w as follows. If the polydisperse sample with composition as listed above would be depolymerized down to its monomer units, and if these units were then reconnected with the premise of now giving the same number of chains as previously (100) but with a monodisperse molar-mass distribution, the result would be a sample with 100 chains of 360,000 g·mol^{-1} each. This new sample would display the same osmotic properties as the original polydisperse one. We may also do the reconnection differently, with the premise of not obtaining 100 but only 66 new monodisperse chains. These chains would then all have a molar mass of 545,000 g·mol^{-1}, and this new monodisperse sample would display the same light-scattering properties as the original polydisperse one.

The preceding example shows us that we receive quite different average values from the same sample, depending on how much emphasis we put on the heavy chains in it. This phenomenon may act as a direct means to express the broadness of the chain-length and molar-mass distribution in just one quantity, which is the **polydispersity index**, defined as PDI = M_w/M_n. In our example above, we have a PDI of 1.51. Commonly, when uncontrolled free-radical chain-growth polymerization or step-growth processes such as polycondensation or polyaddition are employed to synthesize a polymer, a PDI not smaller than 2 can be achieved. This can be calculated from the distribution function that is obtained from such polymerization processes, which is the **Schulz–Zimm distribution**

$$W(P) = \frac{(1-\alpha)^{K+1}}{K!} P^K \alpha^P \tag{1.1}$$

or its variant the **Schulz–Flory distribution** (with $K = 1$)

$$W(P) = P\alpha^P (1-\alpha)^2 \approx P\alpha^P \ln^2\alpha \tag{1.2}$$

Günter Victor Schulz (Figure 8) was born on October 4, 1905, in Łódź, which was then part of the Russian Empire, and moved to Berlin in 1914. He did his undergraduate studies in Freiburg and Munich, working together with Heinrich Wieland and Gustaf Mie, before returning to Berlin for his graduate studies on the thermodynamics of solvation equilibria in colloidal solutions of proteins at the Kaiser Wilhelm Institute of Physical Chemistry and Electrochemistry. He obtained his PhD in 1932, being examined by no other than Fritz Haber on the subject of physical chemistry. He acquired his habilitation in Freiburg under Hermann Staudinger in 1936 and then held an associate professorship at Rostock from 1942, before he was appointed as full professor at the JGU Mainz in 1946, where he established a new Institute of Physical Chemistry out of World War II ruins, which he then headed until his retirement. He died in Mainz on February 25, 1999.

Figure 8: Portrait of Günter V. Schulz. Image reprinted with permission from *Macromol. Chem. Phys.* **2005**, *206*(19), 1913–1914. Copyright 2005 Wiley-VCH.

In eqs. (1.1) and (1.2), $W(P)$ denotes the frequency of occurrence of a certain degree of polymerization P in the sample, in the form of its weight fraction in the sample, $W(P) = m_P/m$. The parameter α denotes the conversion of functional groups (not monomers!) in the case of step-growth processes or the probability for chain propagation relative to the sum of all probabilities (propagation + termination + transfer) for chain-growth processes. The parameter K denotes the degree of coupling, that is, how many individually growing chains form one macromolecule in the end, which is $K = 1$ in the case of termination by disproportionation or $K = 2$ in the case of termination by recombination. In the Schulz–Flory case, with $K = 1$, we get a number-average degree of polymerization of $P_n = 1/(1 - \alpha)$ and a weight-average degree of polymerization of $P_w = (1 + \alpha) / (1 - \alpha)$, and hence, PDI $= P_w/P_n = 1 + \alpha$, which approaches 2 if the reaction approaches full conversion ($\alpha \rightarrow 1$) in the case of step-growth processes. In chain-growth reactions, by contrast, we will get a PDI greater than 2, as in this case, we get a Schulz–Flory distribution for each momentary degree of conversion during the reaction. This is because the parameter α in the Schulz–Flory equation denotes the probability for chain propagation relative to the sum of all probabilities (propagation + termination + transfer) in the case of a chain-growth reaction, and therefore it depends on the monomer concentration, which changes continuously during the process. As a result, the final overall distribution will be an overlay of all these momentary Schulz–Flory distributions, yielding a broad sum distribution with PDI greater than 2. Typically, uncontrolled free-radical polymerization gives PDI = 3. . .10, whereas with the aid of controlling chain-transfer agents, the PDI can be narrowed down to about 2.5 or so. Even narrower distributions with PDI smaller than 2, however, can only be obtained by truly living or quasi-living polymerization processes. In the best controlled case of a living anionic polymerization, the distribution function obtained will be the Poisson distribution

$$W(P) = \frac{\bar{\nu}^{P-1}\, P \exp(-\bar{\nu})}{(P-1)!\;(\bar{\nu}+1)} \tag{1.3}$$

In this equation, $\bar{\nu}$ is the "kinetic chain length", which is the degree of polymerization built from one active, growing molecule before termination; often $\bar{\nu} = P_n$.

With this function, we obtain a PDI of

$$\frac{P_w}{P_n} = 1 + \frac{\bar{\nu}}{(\bar{\nu}+1)^2} \tag{1.4}$$

that approaches 1 with increasing $\bar{\nu} = P_n$. This is because in a living anionic polymerization, a constant number of chains grow steadily. Any initial imbalance of the lengths of these chains will then get less and less significant if they grow longer. As an analogue, consider two kids that are 4 and 2 years old. At that stage, their "age-PDI" is 1.11, but 70 year later, at an age of 74 and 72, their age-PDI is just 1.0002. Nevertheless, though, note that even such a narrow distribution in a polymer sample still has a finite width. For example, consider a sample with number-average degree of polymerization P_n = 500 and P_w/P_n = 1.04. In this sample, still about 32% of the chains have degrees of polymerization of less than 400 or more than 600.

The PDI is related to the general measure of the breadth of a distribution, which is its variance, σ^2, by the relation $\sigma^2 = M_n^2$ (PDI – 1). As a result, the PDI yields information on σ/M_n, but not on σ itself. This means that if two samples have different M_n, the sample with higher PDI may not have the larger σ!

1.3.2.1 Derivation of the Schulz–Flory distribution

Due to its fundamental importance and universal applicability to both chain-growth (i.e., free-radical) and step-growth (i.e., polycondensation and polyaddition) processes of polymer synthesis, the Schulz–Flory and Schulz–Zimm distribution functions, eqs. (1.1) and (1.2), shall be derived in the following. This analytical derivation was originally reported by Schulz for the case of (free-radical) chain-growth polymerization.[3] On top of that, Schulz and Flory were both working on the sister problem of step-growth polycondensation,[4] and a further contribution was made by Zimm even later,[5] but Schulz' work on that matter was first held back by his supervisor Staudinger, so that Flory's report appeared earlier. Both approaches are based on appraising the probability for chain extension steps to occur.

The probability for chain growth in a free-radical polymerization, w_{growth}, which is given by the ratio of the rate of chain growth relative to the sum of the rates of chain growth and termination, is equal to the probability for a certain functional group to have reacted during a step-growth polymerization, α:

$$w_{\text{growth}} = \frac{v_{\text{growth}}}{v_{\text{growth}} + v_{\text{termination}}} = \alpha \tag{1.5}$$

In turn, the probability for chain termination in free-radical polymerization, $w_{\text{termination}}$, and the probability for a certain functional group *not* to have reacted in step-growth polymerization are

$$w_{\text{termination}} = \frac{v_{\text{termination}}}{v_{\text{growth}} + v_{\text{termination}}} = 1 - \alpha \tag{1.6}$$

3 G. V. Schulz, *Z. Phys. Chem.* **1935**, *30B*(1), 379–398.

4 P. J. Flory, *J. Am. Chem. Soc.* **1936**, *58*(10), 1877–1885; G. V. Schulz, *Z. Phys. Chem.* **1938**, *182A*(1), 127–144.

5 B. H. Zimm, *J. Chem. Phys.* **1948**, *16*(12), 1099–1116.

From these two equations, we can derive an expression for the probability of formation of a chain with degree of polymerization P, which we name as w_P. To form such a chain, we need P growth events, the likelihood of which is ${w_{growth}}^P$, and we need one termination event, the likelihood of which is $w_{termination}$:

$$w_P = {w_{growth}}^P \cdot w_{termination} = \alpha^P (1 - \alpha) = \frac{N_P}{N} \tag{1.7}$$

This probability directly translates to the number fraction of chains with degree of polymerization P in the sample, N_P/N, just like the frequency of occurrence of a certain number of points in a dice game with many rolls directly translates to the probability of occurrence of that number in each single dice-rolling event. Note that in the polymer literature, α^{P-1} is sometimes used instead of α^P in eq. (1.7). This is because the start of a polymerization reaction can be seen as R* + M or as R–M* + M, where R is the residue of an initiator, M is the monomer, and * denotes some kind of active species, typically a radical or an ion. However, as P is usually much greater than 1, we may say that $P \approx P - 1$.

The total number of macromolecules in the sample, N, can also be expressed as the total number of chain terminations during the reaction, $n(1 - \alpha)$, because each such event forms one chain:

$$N = n(1 - \alpha) \tag{1.8}$$

where n denotes the total number of all reaction steps. We may limit our focus to cases of α close to 1, as only in that limit, long chains will be present. In the limit of $\alpha \approx 1$, n is equal to the number of *addition reactions*, $n\alpha$:

$$n \approx n\alpha \tag{1.9}$$

From this, it follows that n is practically identical with the number of monomer units in the N macromolecules. The total weight of these N macromolecules, m, in turn, can be expressed as the number of these monomer units times the mass of each unit:

$$m = n \cdot m_{monomer} \tag{1.10}$$

Note that m is not the total weight of the sample, but only that of the N macromolecules in it, without including residual unreacted monomer or other components.

Analogously, the total weight of all macromolecules with the degree of polymerization P is given by

$$m_P = N_P \cdot m_{monomer} \cdot P \tag{1.11}$$

With these two equations, along with eqs. (1.7) and (1.8), we can formulate an expression for the weight fraction of chains with degree of polymerization P

$$\frac{m_P}{m} = \frac{N_P \cdot m_{\text{monomer}} \cdot P}{n \cdot m_{\text{monomer}}} = \frac{N_P \cdot P}{n} = P \cdot \alpha^P (1-\alpha)^2 \tag{1.12a}$$

This is the *Schulz–Flory distribution.* In the polymer literature, a variant form is often found as well:

$$\frac{m_P}{m} = P \cdot \alpha^P \ln^2 \alpha \tag{1.12b}$$

This is because Schulz' original derivation leads to $(m_P/m) = P \cdot \alpha^P (1/\sum P\alpha^P)$. The sum in the denominator is a series of type $\sum P\alpha^P = \alpha + 2\alpha^2 + 3\alpha^3 + 4\alpha^4 + \cdots$. As $\alpha < 1$, this series converges to $1/(1-\alpha)^2$, so that eq. (1.12a) is recovered. Alternatively, if the derivation is conducted in a continuous fashion with the integral $\int P\alpha^P \mathrm{d}P$ instead of the sum $\sum P\alpha^P$, we obtain $1/\ln^2\alpha$ instead of $1/(1-\alpha)^2$,[6] thereby leading to eq. (1.12b).

1.3.2.2 Derivation of the Schulz–Zimm distribution

Let us now examine how such a distribution looks like for the case of *free-radical polymerization with chain termination by recombination.* There are $P/2$ ways for two macroradicals to recombine and form a chain with degree of polymerization P: $1 + (P - 1)$; $2 + (P - 2)$; $3 + (P - 3)$;.........; $P/2 + P/2$.[7] The probability of two macroradicals with degrees of polymerization X and Y to be present at the same time, which is a prerequisite to meet and combine with one another, is

$$w_{X+Y} = 2 \cdot w_X w_Y \tag{1.13}$$

The factor of 2 in eq. (1.13) accounts for the two possible combinations $X + Y$ and $Y + X$. As a result of each such combination, we can obtain a macromolecule with degree of polymerization $P = X + Y$. Rearrangement of that term yields $Y = P - X$, so that with eq. (1.7), we can formulate the likelihood of two macroradicals with degrees of polymerization X and $Y = P - X$ to be present as follows:

$$w_X = \alpha^X (1-\alpha) \tag{1.14a}$$

$$w_{P-X} = \alpha^{P-X} (1-\alpha) \tag{1.14b}$$

This can be inserted into eq. (1.13) to yield

$$w_{X+(P-X)} = w_P = 2 \cdot \alpha^X (1-\alpha) \cdot \alpha^{P-X} (1-\alpha) = 2 \cdot \alpha^P (1-\alpha)^2 \tag{1.15}$$

---

6 Also note that $\ln \alpha = \ln (1 - (1 - \alpha))$, which may be expanded into a Taylor series with a first term of $-(1 - \alpha)$, and so we again get $\ln^2\alpha \approx (1 - \alpha)^2$.

7 This is basically Gauss' math class puzzle, in which he was asked to calculate the sum of all numbers from 1 to 100. Gauss did so by a clever approach: (100 + 1) + (99 + 2) + (98 + 3) +...+ (51 + 50) + (50 + 51), which is 50 × 101. In general: $\sum_{k=1}^{n} k = \frac{n(n+1)}{2}$.

When we multiply this equation with $P/2$, we generate an expression that approximates the number fraction of macroradicals that can yield a chain with degree of polymerization P upon combination:

$$\frac{N_P}{N} = P \cdot \alpha^P (1-\alpha)^2 \tag{1.16}$$

Note that in the literature, α^{P-2} is sometimes used instead of α^P.

From that, we directly get an expression for the number of chains with degree of polymerization P that results from recombination of these macroradicals:

$$N_P = \frac{1}{2} \cdot N \cdot P \cdot \alpha^P (1-\alpha)^2 \tag{1.17}$$

The factor ½ in the latter equation is necessary because recombination of two macroradicals yields only one polymer chain.

By inserting eq. (1.8) into the latter one, we generate

$$N_P = \frac{1}{2} \cdot n \cdot P \cdot \alpha^P (1-\alpha)^3 \tag{1.18}$$

Analogous to eq. (1.12), we can derive an expression for the weight fraction of chains with degree of polymerization P:

$$\frac{m_P}{m} = \frac{1}{2} \cdot P^2 \cdot \alpha^P (1-\alpha)^3 \tag{1.19}$$

In general:

$$W(P) = \frac{(1-\alpha)^{K+1}}{K!} P^K \cdot \alpha^P \tag{1.20}$$

where K is the degree of coupling. Two boundary cases can be distinguished: If $K = 1$, the reaction is only terminated by *disproportionation*, whereas $K = 2$ accounts for termination mostly by recombination.

1.3.2.3 Characteristic averages of the Schulz–Flory and Schulz–Zimm distribution

From eq. (1.7), in its variant with α^{P-1} rather than α^P, we can derive $N_P = N \cdot \alpha^{P-1}(1-\alpha)$, from which we may further derive the number-average degree of polymerization, P_n:

$$P_n = \frac{\sum_{P=1}^{\infty} P \cdot N(P)}{\sum_{P=1}^{\infty} N(P)} = \frac{\sum_{P=1}^{\infty} P \cdot N \cdot \alpha^{P-1}(1-\alpha)}{\sum_{P=1}^{\infty} N \cdot \alpha^{P-1}(1-\alpha)} = \frac{\sum_{P=1}^{\infty} P \cdot \alpha^{P-1}(1-\alpha)}{\sum_{P=1}^{\infty} \alpha^{P-1}(1-\alpha)} \tag{1.21a}$$

where the numerator is the *first* moment of $N(P)$ and the denominator is the *zeroth* moment of $N(P)$.

By factoring out $(1 - \alpha)$, we can write

$$P_n = \frac{1-\alpha}{1-\alpha} \cdot \frac{1\alpha^0 + 2\alpha^1 + 3\alpha^2 + 4\alpha^3 + \cdots}{\alpha^0 + \alpha^1 + \alpha^2 + \alpha^3 + \cdots} = \frac{\frac{1}{(1-\alpha)^2}}{\frac{1}{1-\alpha}} = \frac{1}{1-\alpha} \tag{1.21b}$$

Analogously, from the distribution of weight fractions, $m_P/m = P \cdot \alpha^{P-1}(1-\alpha)^2$, eq. (1.12a), we can derive the weight-average degree of polymerization, P_w:

$$P_w = \frac{\sum_{P=1}^{\infty} P^2 \cdot N(P)}{\sum_{P=1}^{\infty} P \cdot N(P)} = \frac{\sum_{P=1}^{\infty} P^2 \cdot \alpha^{P-1}(1-\alpha)}{\sum_{P=1}^{\infty} P \cdot \alpha^{P-1}(1-\alpha)} \tag{1.22a}$$

where the numerator is the *second* moment of $N(P)$ and the denominator is the *first* moment of $N(P)$.

Again we can factor out $(1 - \alpha)$ to generate

$$P_w = \frac{1-\alpha}{1-\alpha} \cdot \frac{1\alpha^0 + 4\alpha^1 + 9\alpha^2 + 16\alpha^3 + \cdots}{1\alpha^0 + 2\alpha^1 + 3\alpha^2 + 4\alpha^3 + \cdots} = \frac{1+\alpha}{1-\alpha} \tag{1.22b}$$

The PDI, which is the ratio of weight and number-average degree of polymerization, then turns out to be

$$\frac{P_w}{P_n} = 1 + \alpha \overset{\alpha \to 1}{\Rightarrow} 2$$

The derivation for the case of chain termination via recombination, that is, $K = 2$, is analogous to the latter one. It yields

$$P_n = \frac{2}{1-\alpha} \tag{1.23}$$

and

$$P_w = \frac{2+\alpha}{1-\alpha} \tag{1.24}$$

From this, a PDI of 1.5 follows in the limit $\alpha \to 1$

$$\frac{P_w}{P_n} = \frac{2+\alpha}{2} \overset{\alpha \to 1}{\Rightarrow} 1.5 \tag{1.25}$$

We can also calculate the maximum of the Schulz–Flory distribution. We do so for the case of chain termination via disproportionation ($K = 1$):

$$\frac{\partial\left[(1-\alpha)^2 \cdot P \cdot \alpha^P\right]}{\partial P} = (1-\alpha)^2 \cdot P \cdot \alpha^P \ \ln\ \alpha + \alpha^P\ (1-\alpha)^2 \overset{!}{=} 0$$

$$\Leftrightarrow (1-\alpha)^2 \cdot \alpha^{P_{\max}} \cdot (P_{\max} \ln \alpha + 1) = 0$$

$$\Leftrightarrow P_{\max} \ln \alpha + 1 = 0$$

$$\Leftrightarrow P_{\max} = \frac{-1}{\ln \alpha} \approx \frac{1}{1-\alpha} = P_n \tag{1.26}$$

We see that the maximum of the distribution is equal to the number average P_n. The same is true for the case of chain termination via recombination ($K = 2$) (Figure 9):

$$P_{\max} = \frac{2}{1-\alpha} = P_n \tag{1.27}$$

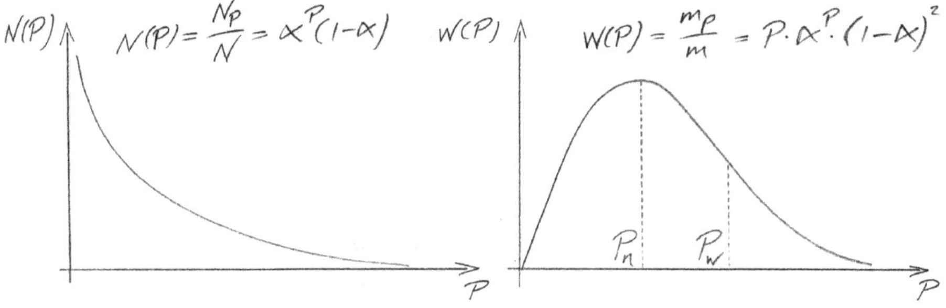

Figure 9: Graphical representation of the number fraction, *N*(*P*), and the weight fraction, *W*(*P*), of each degree of polymerization *P* in a polymer sample according to the Schulz–Flory distribution.

1.4 Properties of polymers in contrast to low molar mass materials

Due to their macromolecular, usually ill-defined structure, polymers exhibit properties that are fundamentally different from their low molar mass counterparts. A comparison of a few such different properties is compiled in Table 1.

Table 1: Comparison of some properties of low molar mass versus polymeric materials.

	Low molar mass materials	*Polymer materials*
Composition	Uniform	Polydisperse and irregular
Solubility	Finite solubility, no swelling	"All or nothing", often with swelling
Melt behavior	Low viscous	Highly viscous or even viscoelastic
Purification	Distillation, crystallization	Precipitation

First, in contrast to low molar mass materials that have a uniform and well-defined molecular constitution (often even in view of their stereochemistry), polymers have a *polydisperse* and irregular *composition*. As detailed in Section 1.3, this is because they are commonly synthesized in statistical chain-growth or step-growth processes, leading to nonuniform chain lengths in a polymer sample, and also leading to irregularities along each individual chain, such as head-to-head versus head-to-tail versus tail-to-tail monomer–monomer connections.

Second, low molar mass compounds can usually be dissolved up to a certain solubility threshold, which is high or low depending on the chemical similarity of the solvent and the compound. This is fundamentally different for polymers: their *solubility* often practically exhibits an "all or nothing behavior", meaning that when a solvent is good it can dissolve virtually any amount of a polymer (at least in the practically relevant concentration range), whereas if a solvent is not good it can hardly dissolve any of the polymer. This is because usually, mixing a compound A (the solute) with a compound B (the solvent) is energetically unfavorable (as A–A and B–B contacts are trivially more similar and therefore more favorable than A–B contacts) but entropically favorable (as entropy generally likes mixing, as that creates disorder). If the compound A has a low molar mass, there's many molecules of A in a given mass of this substance, and hence, the mixing entropy is considerably high, thereby compensating the commonly unfavorable mixing energy and leading to mixing up to a certain threshold concentration at which the solution is saturated. By contrast, if the compound A is a polymer, a given mass amount of it contains just few molecules; hence, the entropy of mixing is so little that it cannot compensate the usually unfavorable energy of mixing, thereby prohibiting dissolution more or less completely. Only if the polymer A and the solvent B are chemically very similar, which is a situation referred to as "good solvent state" (or, even better, "athermal solvent state"), the just little entropy of mixing is still powerful enough to compensate the then also just minorly unfavorable energy of mixing. In this case, the polymer will dissolve in almost any quantities, only limited by practical impairment such as overly high viscosities at high concentrations.

Third, tying in to the preceding sentence, low molar mass molecules usually show low viscosities in the melt and flow like Newtonian fluids (with exemptions such as glycerol or related compounds that have strong mutual intermolecular associative forces causing high viscosities). This is because they are small and can relax (i.e., change their positions relative to each other) comparably fast, without need for input of extraordinarily large additional energies of activation. Polymers, by contrast, are large molecules that need great energies of activation for relaxation and flow, corresponding to *high viscosities*. Especially long-chain molecules can also undergo mechanical entanglement with one another, which imparts further hindrance on their motion. The result is a rich mechanical behavior that is very different from that of small molecules, ranging from elastic snap over viscoelastic creep and relaxation to viscous flow. Whereas the macromolecular chain-like shape of polymers causes this to hold true for all polymers in a universal fashion, the chemical specificity of a given polymer determines what type of mechanical response will be observed at what temperature and on which timescale exactly.

Fourth, polymers do not exist in the vapor phase, because they decompose before boiling (see Figure 3). Therefore, standard *purification* procedures such as distillation cannot be applied for polymers. Fortunately, however, tying in to our second point above, their "all or nothing" solubility allows polymers to be precipitated from a solution by adding an excess of a nonsolvent, thereby keeping low molar mass impurities dissolved while the polymer separates from the solution and can be harvested just by filtration.

1.5 Polymers as soft matter

Polymers are a prime example of a material class named **soft matter**. This term refers to materials that (i) consist of building blocks with sizes in the **colloidal domain**, 10–1000 nm, and that (ii) have interaction energies in the range of 10–100 RT (as a benchmark to that, note that RT at room temperature is about $8\ \mathrm{J\ mol^{-1}\ K^{-1}} \cdot 300\ \mathrm{K} \approx 2.5\ \mathrm{kJ\ mol^{-1}}$).[8] To put this into context, let us consider that a covalent C–C bond has a dissociation energy of 350 kJ mol^{-1}, which is more than 100 RT at room temperature; this means that it requires a huge amount of external energy to be broken. A transient OH···H interaction, by contrast, has a dissociation energy of just 18 kJ mol^{-1}, and can therefore be broken with just little additional energy. Together, the large size and weak interactions of the building blocks in soft matter make it soft. We can appraise that by estimating the cohesive energy density, $e = E/r^3$, where E is the interaction energy and r is the distance between the building blocks. For hard matter, we have E in the range of 10^{-18} J and r in the range of 0.1 nm, which results in values of e around 10^{12} Nm^{-2}; this is a typical elastic modulus for materials such as diamond. For soft matter, in

8 A little footnote in this context: the value of RT at room temperature should be in the mind of each physical chemist. This is because it is the fundamental benchmark of the energy landscape, which is a land that physical chemists often have to pass. As a comparison, consider the following analogy: an information like "the building over there is 10 m tall" is only valuable for you if you know what 1 m is. If you do not know that, then you have no clue whether 10 m is tall or short. The same applies to energies, with which you need to argue over and over again in physical chemistry. You can only appraise whether a 350 kJ mol^{-1} covalent C–C bond or a 20 kJ mol^{-1} hydrogen bonding interaction is strong or weak if you know the baseline of RT at room temperature.

contrast, we have E in the range of 10^{-20} J and r in the range of 10–100 nm, which results in values of e around 10^4–10^7 Nm^{-2}. This is a typical elastic modulus for a polymeric or colloidal melt or gel. The weak interaction energy also gives rise to another characteristic property of soft matter: its ability to be assembled, disassembled, and reassembled in a dynamic manner, thereby displaying a rich dynamic and stimuli-sensitive phase behavior. Due to that and due to their colloidal-scale size, these materials (iii) have **relaxation times**, which are the times needed for their building blocks to move a distance of their own size, in the practically relevant range of milliseconds to seconds. As a result, soft matter materials often show rich **viscoelasticity**, depending on the timescale of experimentation and temperature.

Following this definition, polymers can be classified as a prime example of soft matter. First, the shape of a polymer chain is, in most cases, that of a random coil with a couple of 10 nm in diameter. This puts polymers right in the colloidal domain. Second, due to their colloidal-scale size, the polymeric coils exhibit relaxation times in the range of a couple milliseconds to seconds.[9] Third, multiple polymer chains can interact with each other usually via van der Waals or dipole–dipole interactions to form higher-order assemblies. These interactions have energies in the range of a couple 10 RT. Consequently, polymeric assemblies are stable at room temperature, but can be restructured with relative ease, which is one out of two reasons why nature often uses polymeric structures as buildings blocks for *dynamic* complex architectures. (The second reason will be discussed in the context of the Flory–Huggins theory in Chapter 4.) Aside from this fundamental role in nature, the soft-matter characteristics of polymers have made tremendous impact on science. The basic physics of ordering phenomena is the same for colloidal-scale soft matter and for classical atomic- or small-molecule-scale hard matter, but soft matter, due to the much larger size and slower motion of its building blocks compared to that of atoms and small molecules, can be powerfully observed and studied. This notion led Pierre Gilles de Gennes to use polymers as fundamental studying objects in condensed-matter physics, which is much more elementary than "only" studying them for their vast practical applicability. This groundshifting inspiration was honored with the Physics Nobel Prize in 1991.

9 The exact time depends on the architecture and the environment of the polymer chain. A short chain will relax faster than a long one. It will also relax faster when it is surrounded only by solvent molecules, as it would be the case in a dilute polymer solution, than when it is in direct contact with other polymer chains that constrain its movement, as it would be the case in a polymer melt.

Pierre Gilles de Gennes (Figure 10) was born on October 24, 1932, in Paris. He was home-schooled to the age of 12. By the age of 13, he had adopted adult reading habits and was visiting museums. Later on, he studied at the École Normale Supérieure, from which he majored in 1955. From 1955 to 1959, he was a research engineer at the Atomic Energy Center (Saclay), the heart of French nuclear research at that time, and obtained his PhD in 1957 from the University of Paris. After postdoctoral research in Berkeley in 1959 and a 27-month stint in the French Navy, he became an assistant professor in Orsay in 1961. Later, in 1971, he became full professor at the Collège de France. From 1976 to 2002, he was the director of the Ecole de Physique et Chimie. After receiving the Physics Nobel Prize in 1991, de Gennes decided to give talks on science, innovation, and common sense to high school students. He died in Orsay on May 18, 2007.

Figure 10: Portrait of Pierre Gilles de Gennes. Image reproduced with permission from *Nature* **2007**, *448*(7150), 149. Copyright 2007 Springer Nature.

1.6 History of polymer science

For centuries, natural polymers have been employed without any knowledge about their macromolecular nature. Table 2 shows a brief history of mankind's use of them.

Table 2: Use of natural polymers in human history.

Time	Material	Discovery
5000 BC	Cotton	Mexico
3000 BC	Silk	China
2000 BC	Bitumen	Orient
1500 AD	Rubber	Columbus
1800 AD	Guttapercha (*trans*-1,4-isoprene)[1]	
1839 AD	Vulcanized rubber (*cis*-1,4-isoprene)	Goodyear

[1]*Gutta* "rubber" and *percha* "tree".

In this list, Goodyear's vulcanization of natural rubber is a special milestone, because it is the first industrial-scale modification of a natural polymeric material, and therefore signifies the advent of a chemical industry in this field. From thereon, chemists modified many biopolymers, but still had no knowledge about their structure or their macromolecular nature. It was believed at that time that covalently jointed structures with molar masses of several thousand grams per mole simply could not exist. Instead, a common belief was that polymers were colloidal structures, built up by small molecular entities aggregated by noncovalent forces. The reason for this notion might have been because organic chemistry was successful in assessing small molecule structures to that time, whereas physical chemistry was successful in assessing their intermolecular forces, so that scientists were prone to apply both polymer-type matters as well. The high molar masses of these materials that were measured in laboratories were therefore viewed to be that of the colloidal aggregates, and hence, sort of artifacts. Table 3 displays the hypothetic and actual molecular structures of caoutchouc and cellulose as they were perceived before 1920 and as we know them today.

Table 3: "Colloidal" and actual structure of caotchouc and cellulose.

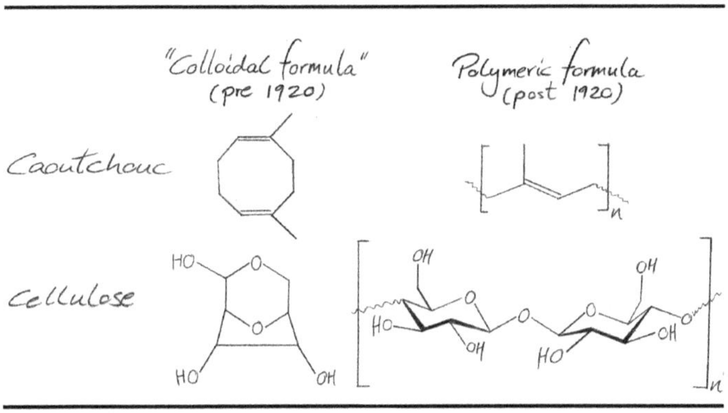

	"Colloidal formula" (pre 1920)	Polymeric formula (post 1920)
Caoutchouc		
Cellulose		

In 1920, Hermann **Staudinger** was able to prove that polymers, in fact, have actual rather than artifactual high molar masses. He did this by chemically modifying known polymers, as shown in Figure 11, with the premise that such chemical change in their "colloidal" structures should severely alter their aggregation, just in the same way that a colloidal fatty acid micelle breaks up upon esterification. What Staudinger found instead was staggering: despite drastic changes in many other properties, the high molar masses were retained in all the compounds upon such modification (which has therefore been referred to as "polymer analogous modification" in the following). This finding strongly supports Staudinger's hypothesis that polymers are not associated by physical interactions only, but instead, covalently jointed chain molecules with an inherently high molar mass. This mass doesn't change upon polymer analogous modification, whereas other properties do, as the modification alters the side groups of the polymer, but not the length of its chain backbone.

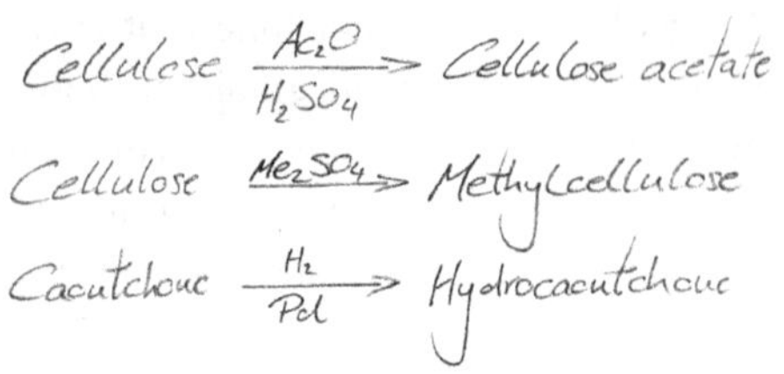

Cellulose $\xrightarrow[H_2SO_4]{Ac_2O}$ Cellulose acetate

Cellulose $\xrightarrow{Me_2SO_4}$ Methylcellulose

Caoutchouc $\xrightarrow[Pd]{H_2}$ Hydrocaoutchouc

Figure 11: Hermann Staudinger's polymer analogous reactions that proved a high molar mass to be an actual rather than just artificial polymeric property.

Staudinger expressed his findings rather carefully first:

> If one is willing to imagine the formation and constitution of such high-molecular compounds, one may assume that primarily, a combination of unsaturated molecules has taken place, similar to the formation of four- or six-membered rings, but that for some reason, potentially a steric one, the four- or six-ring closure didn't happen, and now many, perhaps hundreds of molecules are jointed.[10]

Staudinger's notion received immediate and harsh criticism by the scientific community of the early twentieth century, such as the following, which was expressed by Heinrich Wieland, his predecessor at the University of Freiburg and Nobel Prize laureate for the structural characterization of bile acids:

> Dear colleague, refrain from the imagination of large molecules; organic molecules with a molecular weight above 5000 do not exist. Purify your products, such as the caoutchouc, and then they will crystallize and turn out to be low-molecular compounds.[11]

But Staudinger proved persistent, and, as his hypothesis held true, grew increasingly self-confident later, when he said:

> Even though the formation of millions of compounds whose molecules are built of hundred or several hundreds of atoms is possible, this realm of low-molecular organic chemistry is just a pre-stage to actual organic chemistry, namely chemistry of compounds that are of fundamental importance for processes in life. This is because the latter compounds contain molecules that

10 The original German wording is: "Will man sich eine Vorstellung über die Bildung und Konstitution solcher hochmolekularen Stoffe machen, so kann man annehmen, daß primär eine Vereinigung von ungesättigten Molekülen eingetreten ist, ähnlich einer Bildung von Vier- und Sechsringen, daß aber aus irgendeinem, evtl. sterischen, Grunde der Vier- oder Sechsringschluß nicht stattfand, und nun zahlreiche, evtl. hunderte von Molekülen sich zusammenlagern."

11 The original German wording is: "Lieber Herr Kollege, lassen Sie doch die Vorstellung mit den großen Molekülen, organische Moleküle mit einem Molekulargewicht über 5000 gibt es nicht. Reinigen Sie Ihre Produkte, wie z.B. den Kautschuk, dann werden diese kristallisieren und sich als niedermolekulare Stoffe erweisen."

> are not built from some hundreds, but from thousands, ten-thousands, and perhaps even millions of atoms.[12]

Staudinger eventually received the Nobel Prize for his work in 1953. By then, the field of macromolecular chemistry that he founded had advanced impressively, and also his above statement about polymers as the "true organic molecules" and backbones of life held true, as in the same year, Watson and Crick published the structure of DNA.

Hermann Staudinger (Figure 12) was born on March 23, 1881, in Worms. His father, a leading figure in the German union movement, wanted his son to enter a hands-on profession, so Hermann became an apprentice in carpentry first. After his training, he went to Halle to study chemistry and obtained his PhD there. He then held professorships in Karlsruhe, Zurich, and finally at Freiburg, where he conducted his Nobel Prize research. He died in Freiburg on September 8, 1965.

Figure 12: Portrait of Hermann Staudinger. Image reproduced with permission from ETH-Bibliothek Zürich, Bildarchiv.

In the 40 years following Staudinger's groundbreaking experiments (1920–1960), the core concepts of polymer physics were developed:
- Kuhn model for polymer chain renormalization (Chapter 2)
- Flory theory of coil conformation in a good solvent (Chapter 3)
- Rouse and Zimm models of polymer dynamics (Chapter 3)
- Flory–Huggins mean-field theory of polymer thermodynamics (Chapter 4)
- Kuhn, Grün, Guth, Mark: statistical theory of rubber elasticity (Chapter 5)

Between 1960 and 1980, these concepts were further developed and refined to move from the description of single chains to chain assemblies, culminating in the Nobel

---

12 The original German wording is: "Wenn auch die Darstellung von Millionen von Stoffen möglich ist, deren Moleküle aus Hundert oder einigen Hundert Atomen aufgebaut sind, so bliebe dieses Gebiet der niedermolekularen organischen Chemie trotzdem nur die Vorstufe der eigentlichen organischen Chemie, nämlich der Chemie der Stoffe, die für die Lebensprozesse von grundlegender Bedeutung sind. Denn in letzteren Verbindungen liegen Moleküle vor, die nicht aus einigen Hundert, sondern aus Tausenden, Zehntausenden und vielleicht sogar Millionen von Atomen bestehen."

Prize for Pierre Gilles de Gennes in 1991 for discovering that methods developed for studying order phenomena in simple systems can be generalized to more complex forms of matter, in particular, to liquid crystals and polymers. We just name two bullet points to summarize these milestones:
- Doi, Edwards, De Gennes: tube concept and reptation theory (Chapter 5)
- De Gennes: "soft matter physics"

Current topics of polymer research (with a special focus on the physical chemistry branch of it) have introduced dynamics and adaptivity into polymer systems by actively incorporating or making use of transient bonding. This leads to **self-assembly** processes that create complex architectures potentially soon rivaling those in nature; it also leads to **sensitivity**, responsiveness, and adaptivity of these complex systems and materials. To freely quote Bruno Vollmert's book Grundriss der Makromolekularen Chemie: "Polymers are the highest level of complexity in chemistry, and the lowest level of complexity in biology." Current polymer research focuses right on the bridge between these realms. Other modern topics include:
- Supramolecular polymers (J. M. Lehn, E. W. Meijer, T. Aida, M. A. Cohen-Stuart, and many others)
- Conductive polymers and polymers for energy conversion and storage
- Biopolymer science and biomimetics
- Large-scale theoretical modeling and simulation

Aside and along with this academic development, industrial polymers have turned mankind's twentieth century into the "polymer age".[13] This vast and versatile use of polymers throughout decades, however, has posed environmental challenges that modern polymer research must address, in the form of reversing the tremendous amounts of polymer-based waste that has been and still is being produced by mankind. Modern research, both industrial and academic, therefore must have a special focus on **degradable** polymeric materials. On top of that, it will be a sociological challenge for us to learn about the great value of polymer-based everyday life products, given their remarkable properties (mostly their strong mechanics compared to their lightweight), such as to not only use them once but multiple times before discarding them.

13 Just as much longer before, mankind's main materials of use led us to entitle former ages as the stone age, the bronze age, and the iron age, and just as our world today is certainly justified to be termed to be in the silicon age.

2 Ideal polymer chains

Lesson 2: Ideal Chains

In your elementary physical chemistry classes, the first and most simple type of matter that you dealt with was the ideal gas. In polymer science, this has an analog. The first and most simple type of polymeric matter that we will deal with is the ideal chain. In this lesson, you will get to know the basics of the ideal chain model, see how this is conceptually very related to the ideal gas analog, and you will see how this model allows us to simply capture and quantify the structure of a polymer.

The goal of this book is to establish relations between the *structure* and *properties* of polymers. To do so, we have to find a way to comprehensively describe the structure of the elementary building block of each polymeric material: the single polymer chain. Polymers are made from a large number of repeating units; as such, a polymer chain is a complex, multibodied entity that is hard to describe analytically. We therefore have to rely on models that simplify the complexity while simultaneously retaining the physical realities that we observe in experiments. In the following, we will introduce a few of these models, starting with the simplest one: the **ideal polymer chain**.

Let us recall a concept that you have encountered in the beginning of your basic physical chemistry lectures, where you have considered the most simple state of matter imaginable: the *ideal gas*. The description of an ideal gas is based on the assumptions that (i) the gas molecules are point-like particles with no volume and (ii) no interactions (other than simple elastic collisions). These assumptions are, of course, not true: real gas molecules do have a finite volume, and they do interact with each other in many ways, especially at short mutual distances, which is given at high pressure. In many cases, however (for instance, at usual temperature[14] and pressure), gases in fact widely behave ideal, allowing us to use the above very simplified model to treat them, which gives us (iii) a very simple equation of state: $pV = nRT$, with p the pressure, V the volume, n the molar number of gas particles (atoms or molecules) in our system, R the gas constant, and T the absolute temperature. The concept of the ideal polymer chain is based on the same simplistic principle: it imagines the chain to consist of (i) monomer segments with no volume and (ii) no interactions with one another or the environment (other than the trivial interaction that they have due to their mutual connectivity, which, as we will see in Chapter 3, impairs their independent statistical motion). The model of the ideal polymer chain is the basis for any further development of refined polymer chain models and can therefore rightfully be called a starting point of polymer physics.

14 Strictly speaking: at a temperature well above the critical one.

https://doi.org/10.1515/9783110672817-002

We will see later in this chapter that this model will give us (iii) an equation of state that is strikingly similar to the ideal gas law, both in view of its mathematical appearance and in view of its simplicity but yet powerful utility.

2.1 Coil conformation

2.1.1 Micro- and macroconformations of a polymer chain

Let us begin by considering how a polymer chain might look like. The shape of a polymer is determined by three factors, as illustrated in Figure 13(A). The first factor is the monomer segmental length, that is, the length of the elementary structural unit plus the bond that connects it to the next. In many simple vinyl polymers, such as polyethylene (PE), polypropylene (PP), polyvinylchloride, or polystyrene (PS), this is a covalent carbon–carbon bond with a length of 1.54 Å. The second factor is the monomer bonding angle, φ. It has a value of 109.6° for an sp^3-hybridized carbon–carbon bond. Both of these parameters are determined by the specific chemistry of the monomer, which means that they are fixed for a given polymer. The third factor is the monomer bond torsion angle, θ. This factor is not fixed, because the bonds can rotate quite easily. As a consequence, the macroconformation of a chain will result from the sequence of its monomer bond torsional microconformations. A polymer chain may therefore adopt many different shapes, two of which are depicted in Figure 13(B) and (C). In Figure 13(B), you see a structure in which all the bonds are *trans*-conformed. This all-*trans* conformation leads to an ordered, rod-like macroconformation. Entropy, however, does not favor such ordered states. More favorable, by contrast, is the structure depicted in Figure 13(C). Here, the microconformation is a random *cis–trans* mix that leads to a macroconformation of a random coil, which does not show a high degree of order. The actual reason for the entropic favor of this random-coil macroconformation is because it is realized by many different microconformations of kind as the one shown in Figure 13(C), because there are many different possibilities for arrangement of a random *cis–trans* mix along the chain contour, whereas the rod macroconformation is realized by just one microconformation, namely the all-*trans* chain. With that, the core principle of statistical thermodynamics applies: a macrostate that is realized by a high number of microstates has a higher entropy and is more likely to be observed (both over time and ensemble) than a macrostate that is realized by just a few microstates.

So far, we have discussed the monomer bond torsional angle θ and the resulting structure of a polymer chain primarily from an entropic perspective. There is, however, also an energy side to the argument. This is because some monomer bond torsional angles are energetically favorable, whereas others are not. The *trans* or *gauche* conformations have lower free energies, ΔE, than the *cis* or the *anti* ones, as shown in Figure 14 for the very typical case of a carbon main chain. The energy

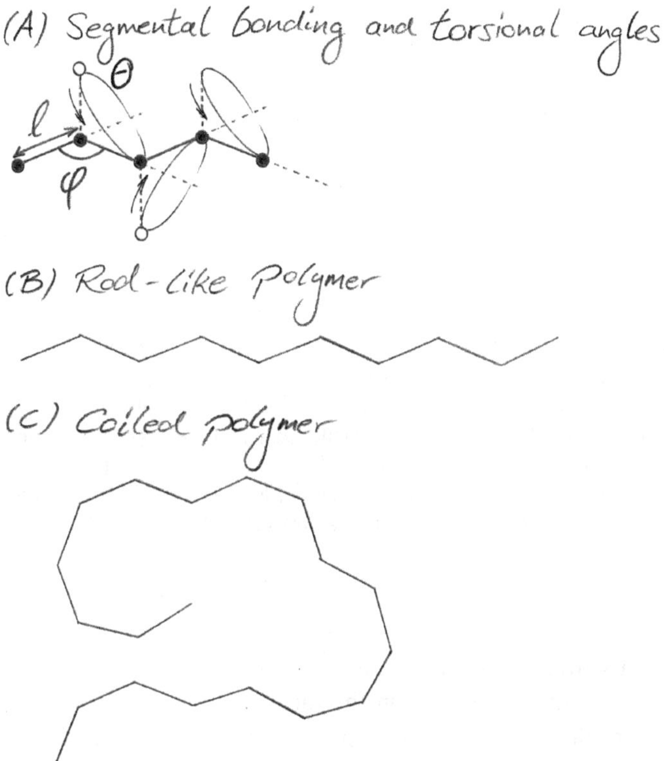

Figure 13: (A) Characteristic quantities of the elementary chemical bonds in the backbone of a polyethylene chain. In this simplest case of a polymer, the bond length between two methylene units, here denoted as dark-shaded dots, corresponds to the segmental length, l. The monomer bonding angle, φ, is the angle between two segment–segment (here: methylene–methylene) bonds, and the monomer bond torsion angle, θ, defines the conformational positions of each unit. Picture taken from H. G. Elias: *Makromoleküle, Bd. 1 – Chemische Struktur und Synthesen* (6. Ed.), Wiley VCH, **1999**. Depending on the angle θ, many different macroconformations of the polymer chain can be realized, two of which are shown here. **(B)** If all microconformations of the chain are *trans*, this leads to an ordered, rod-like macroconformation that is entropically unfavorable. **(C)** In contrast, if a more favorable random microconformational mix of *cis* and *trans* is present, this leads to a macroconformation of a random coil. This unordered structure is favored by entropy, because it is realized by more microconformations than the all-*trans* structure in (B).

difference between those conformations, however, is just small: the energy of the *trans* or *gauche* conformations differs by only 3 kJ mol^{-1}, which corresponds to just about 1.2 RT at room temperature. Furthermore, a conformational change between these states has an energy of activation of only 13 kJ mol^{-1}, that is, just about 5 RT. Even the biggest energy gap, which is between the *cis* and the *trans* conformation, is only about 17 kJ mol^{-1}, that is, 7 RT. This means that all the conformations available to the building blocks of a polymer chain are not very different in energy.

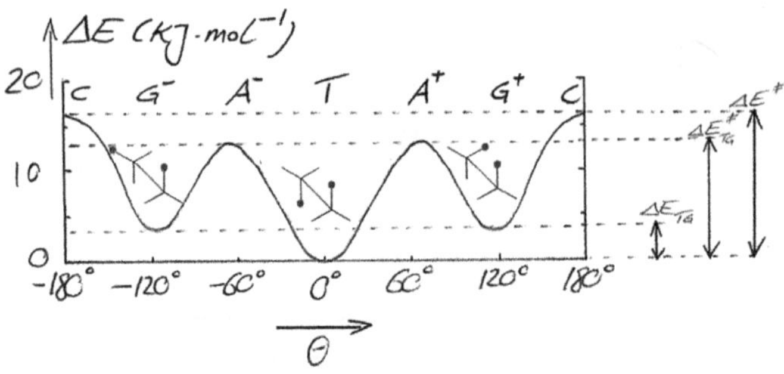

Figure 14: Dependence of the molar energy, ΔE, on the bond torsion angle, θ, for basic carbohydrates. Conformations are abbreviated as C = *cis*, G = *gauche*, A = *anti*, T = *trans*. The energy has an absolute minimum at the *trans* conformation and two further local minima for the *gauche* conformations G^+ and G^- that are 3 kJ mol^{-1} higher (ΔE_{TG}). The energy of activation needed to change between T and G^+ or G^- is 13 kJ mol^{-1} ($\Delta E^{\ddagger}_{TG}$), and the energy gap between the most favorable and the least favorable conformation, T and C, is 17 kJ mol^{-1} ($\Delta E^{\ddagger}$). Picture redrawn from H. G. Elias: *Makromoleküle, Bd. 1 – Chemische Struktur und Synthesen* (6. Ed.), Wiley VCH, **1999**.

Moreover, it is easy to pass from one conformation to the other, because a conformational change needs an activation energy in the range of only a couple RT at room temperature. As a result, these changes happen fast, which is the prime reason why polymers are **flexible** above their glass transition temperature. This flexibility is lost if there is a more severe energy barrier between the monomer segmental conformations, which is the case for monomer units that carry bulky substituents. This is the reason why PS, which has a bulky benzene side group, has a higher glass transition temperature than PE, where no side groups are present, because it takes more thermal energy to activate conformational change dynamics in the more obstructed PS backbone than in the flexible PE backbone. (See: here we have our very first qualitative structure–property relation!)

As another direct consequence of these low energy values, the entropy aspect is mainly responsible for the macroconformation of a polymer chain and outweighs the energy aspect. Therefore, **the most probable structure of a polymer chain is that of a flexible random coil.** There are some cases where energy outweighs entropy, but these only occur when an extremely high energy gain is associated with the adoption of a certain specific chain structure. Proteins are a prime example: they have specific, energetically favorable folded shapes, such as the α-helix or the β-sheet. The energy they gain through secondary interactions when adopting these shapes outweighs the entropy penalty that such an ordered structure imposes. This ordered structure, in turn, is the basis for their biological function, often referred to as key-and-lock principle. This notion has led Vollmert to formulate that biopolymers such as proteins are the highest level of complexity in chemistry and the lowest level of complexity in biology. Most synthetic polymers, by contrast, do not have such preferred structures, but instead shape up as random coils.

Now that we know that a polymer chain looks like a coil, how do we describe it mathematically? Even if we restrict ourselves to just the energetically favored *trans* and *gauch* microconformations, T, G^+, and G^-, there are 3^{N-2} possible macroconformations that result in a chain with N segments. If we consider N to be 100, which is still a comparably short polymer chain, that would mean that there are 6×10^{46} possible different macroconformations! On top of that, conformational change is very fast due to the low energies involved. An appraisal based on an Eyring-type equation with an energy barrier of 13 kJ mol^{-1}, or 5 RT at room temperature, for the T $\rightarrow$ G transition shows that they happen on a timescale of just nanoseconds at room temperature. All this means that the shape of a polymer chain cannot be described analytically. It can, however, be described mathematically using suitable *average* values.

2.1.2 Measures of size of a polymer coil

Two common averages are used to describe the size of a polymer coil. These are the **end-to-end distance**, $\vec{r}$, and the **radius of gyration**, R_g, both depicted in Figure 15. The end-to-end distance, $\vec{r}$, is calculated by simply summing up all individual bond vectors, which creates a vector that points from one end of the polymer chain to the other:

$$\vec{r} = \vec{r}_1 + \vec{r}_2 + \vec{r}_3 + \cdots = \sum \vec{r}_i \tag{2.1}$$

It can be easily visualized and imagined, but it is hard to be determined experimentally, which is why it is often preferred by polymer theorists. Note that the end-to-end distance is only meaningful for linear chains, which have a defined start and end of their polymer chain. Branched macromolecules, by contrast, have many end-to-end distances, because they have multiple chain ends.

The radius of gyration, R_g, is calculated by the sum over all mass segments, m, weighted by their distances to the center of mass:

$$R_g^{\ 2} = \frac{1}{m}\left(\vec{r}_1^{\ 2} m_1 + \vec{r}_2^{\ 2} m_2 + \vec{r}_3^{\ 2} m_3 + \cdots\right) = \frac{1}{m}\sum \vec{r}_i^{\ 2} m_i = \frac{1}{m}\int r^2 \mathrm{d}m \tag{2.2}$$

R_g corresponds to about the 1.3-fold of the radius of a sphere with the same moment of inertia and the same density as our actual object of interest. It is not limited to linear polymers, but can also be determined for branched or cross-linked polymers. In fact, R_g can be calculated for any geometrical object. It is also much easier to assess experimentally than the end-to-end distance, which is the reason why it is often preferred by polymer experimentalists.

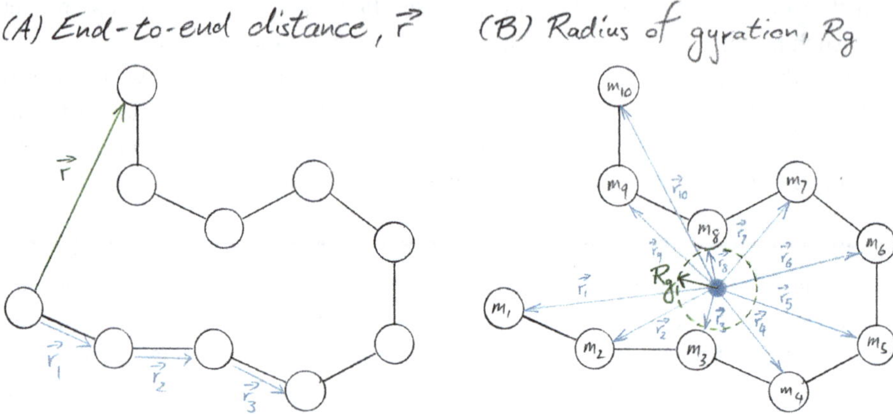

Figure 15: Schematic representation of two common average measures of the size of a polymer coil: **(A)** The *end-to-end distance* $\vec{r}$, simply calculated as the sum of all bond vectors, and **(B)** the *radius of gyration*, R_g, calculated by the sum over all mass segments weighted by their distances to the center of mass of the polymer coil; R_g corresponds to the radius of a solid sphere (illustrated by a dashed green circle) with the same moment of inertia and the same density as the fuzzy polymer coil.

Both quantities are connected to each other by simple relations that depend on the geometry of the compound investigated. Table 4 compiles a selection of some of these relations.

Table 4: Relation between the end-to-end distance and the radius of gyration for various types of objects.

Type of molecule/ object	$\langle R_g^2 \rangle$ – characteristic length	Characteristic length–mass	
Random coil	$\langle R_g^2 \rangle = \frac{\langle \vec{r}^2 \rangle}{6}$	$\langle \vec{r}^2 \rangle \sim \langle R_g^2 \rangle \sim m$	
Coil in good solvent	$\langle R_g^2 \rangle = \frac{\langle \vec{r}^2 \rangle}{(2\nu+1)(2\nu+2)}$	$\langle \vec{r}^2 \rangle \sim R_g^2 \sim m^{2\nu}$	ν = Flory exponent (Chapter 3) → ideal coils: 0.5 → with rep. interact: 0.6
Rod with length L	$\langle R_g^2 \rangle = \frac{L^2}{12}$	$L \sim m$	
Disc with radius r	$\langle R_g^2 \rangle = \frac{r^2}{2}$	$r \sim m^{1/2}$	
Sphere with radius r	$\langle R_g^2 \rangle = \frac{3r^2}{5}$	$r \sim m^{1/3}$	

2.2 Simple chain models

2.2.1 The random chain (phantom chain, freely jointed chain)

The simplest model to describe a polymer chain is that of the **random chain**, also referred to as *freely jointed chain* or *phantom chain*. It is based on the assumption that all monomer bond conformations, assessed by the torsional angle, θ, and also even all monomer–monomer bonding angles, φ, are possible. This is somewhat unintuitive, as the bonding angle is in fact set by the monomer's specific chemistry. Releasing this constraint and allowing this angle to be free would allow two monomer segments to occupy the same spot in space, for example, if two adjacent monomer segments would have a bonding angle of $\varphi = 0°$, which is, nevertheless, allowed to be possible in the random chain model. This is why this model is also referred to as the *phantom chain model*, as in this framework, the chain segments can penetrate through each other like phantoms. With that simplifying assumption, the random chain model relates the number of monomer segments, N, and their length, l, to the end-to-end distance of the polymer chain, $\vec{r}$, in a very simple fashion, as we will show now by simply estimating the end-to-end distance through the bond vector sum. For a reason that we will detail below, we use the *square value* of the end-to-end vector, $\vec{r}^2$. With that, we calculate the square of the bond vector sum using a scalar product term:

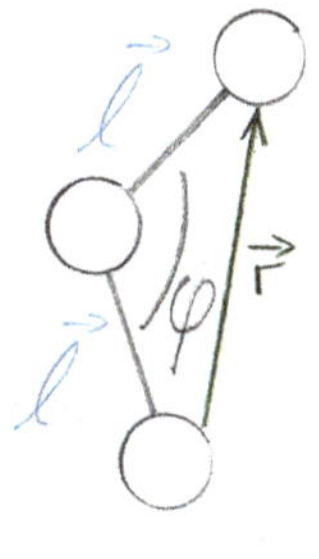

$$\vec{r} = \sum \vec{l}_i \tag{2.3a}$$

$$\vec{r}^2 = \left(\vec{l}_1 + \vec{l}_2 + \vec{l}_3 + \cdots\right)^2$$
$$= \left(l_1^{\,2} + l_2^{\,2} + l_3^{\,2} + \cdots\right) + \sum l_i l_j \cos\varphi_{ij} \tag{2.3b}$$

With this calculation, so far, we have only estimated the *momentary*-squared end-to-end distance of *one single* polymer coil. This, however, is not representative, as the coil conformation is subject to constant dynamic change, and as in a sample with many chains, even in a momentary view, they won't all exhibit the same end-to-end distance, but a *distribution* of it. So, what we are actually interested in is the *mean* of this distribution. If we would calculate it for $\vec{r}$ directly, however, the mean would always be zero. This is because $\vec{r}$ is a vector quantity that has a direction. In a sample with many representants of this quantity, there will always be pairs of same length that point into opposite directions, such that the mean of all these would cancel out to zero. To get rid of this circumstance, we take the square of the end-to-end vector first before averaging. That way, we eliminate the directional dependence and obtain the **mean-square end-to-end distance:**

$$\begin{aligned}\langle \vec{r}^2 \rangle &= \left(l_1{}^2 + l_2{}^2 + l_3{}^2 + \cdots\right) + \sum l_i l_j \langle \cos \varphi_{ij} \rangle \\ &= \left(l_1{}^2 + l_2{}^2 + l_3{}^2 + \cdots\right) \\ &= Nl^2 \end{aligned} \tag{2.4}$$

In the latter form of the scalar product, it is very convenient that the average of the angular dependence, $\langle \cos \varphi_{ij} \rangle$, is zero due to the assumption that the bonding angle, φ, is free and exhibits a random distribution. The square of the end-to-end distance therefore only depends on the number of monomer segments, N, and their length, l.

With this calculation, we have derived a simple **power law** for the chain mean-square end-to-end distance as a function of the number of chain segments. To re-linearize the physical dimension of the mean-square end-to-end distance, we can take the square root and obtain the **root-mean-square** (rms) **end-to-end distance**, $R = \langle \vec{r}^2 \rangle^{1/2}$. Such rms values can be found all over the field of physical chemistry, as they are a practical tool to handle vector quantities that are subject to a distribution. rms values eliminate the vectors' directional dependencies and handle their distributed nature by using the arithmetic mean of their square values. From this, the square root is then formed to re-linearize the mean square quantity to its truly meaningful physical dimensions. Following eq. (2.4), the rms end-to-end distance of our phantom polymer chain scales with $N^{1/2}$:

$$R = \langle \vec{r}^2 \rangle^{1/2} \sim N^{1/2} \tag{2.5}$$

Note that from eq. (2.5), it follows that a polymer that has twice as many segments as another one is not twice as large as that other one in space, but only $2^{1/2} \approx 1.4$ times as large. To get a polymer that is twice as large as another one in space, it must have $2^2 = 4$ times as many segments!

A very similar scaling law to eq. (2.5) can be found for freely diffusing particles that move by a **random walk**. Here, the rms displacement, $\langle \vec{x}^2 \rangle^{1/2}$, scales with the square root of the number of diffusion steps, and therefore also with the square root of time, $t^{1/2}$, if we assume that each elementary step takes a defined period. This scaling is based on the Einstein–Smoluchowski equation:

$$\langle \vec{x}^2 \rangle^{1/2} \sim t^{1/2} \tag{2.6}$$

The similarity of eq. (2.5) and eq. (2.6) is striking. Upon closer scrutiny, however, it is reasonable. The shape of a random phantom coil composed of bonds with fixed length l and free random bonding angles φ is the same as that of a path of diffusion steps of fixed length l and free random directional change from step to step, as shown in Figure 16.

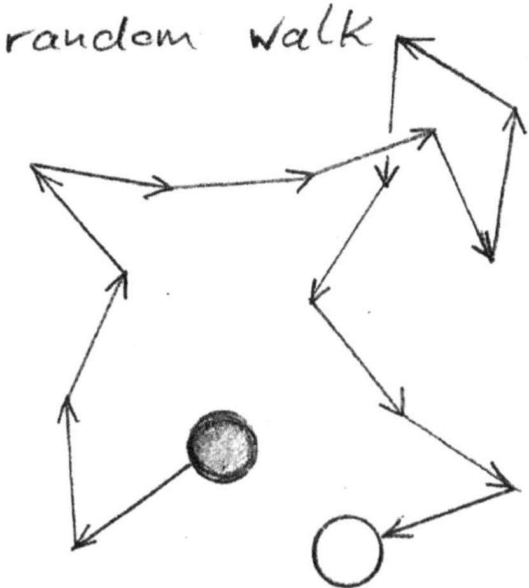

Figure 16: Trajectory of a two-dimensional random walk.

2.2.2 The freely rotating chain

So far, we have assumed no angular dependence in our calculations. However, we actually know how such a dependence should be described, because we know the bonding angle, φ, from the chemical specificity of each monomer. Thus, we can assume that each bond vector $\vec{l}_j$ projects a component $l \cdot \cos\varphi$ onto the next bond vector $\vec{l}_{j+1}$. Accounting for that, eq. (2.4) can be refined to

$$\langle \vec{r}^2 \rangle = Nl^2 \frac{1 - \cos\varphi}{1 + \cos\varphi} \tag{2.7}$$

This equation is derived through a series expansion, which is explained in detail in Rubinstein's and Colby's book *Polymer Physics*. When we calculate $(1 - \cos\varphi)/(1 + \cos\varphi)$ for an sp^3-hybridized carbon bond with a bonding angle of $\varphi = 109.6°$, we get a value of 2. Consequently, we see that we underestimate such a polymer coil's size by a factor of $\sqrt{2}$ if we disregard its fixed bonding angle, as we do in the phantom chain model.

2.2.3 The chain with hindered rotation

We can further take into account the bond's torsional angle, θ. We have seen in Section 2.1 that the torsional angle is not fixed, but prefers specific conformations due to their lower energies relative to other conformations. Accordingly, we can average over all possible bond conformations per time and chain ensemble to determine the average torsion angle $\langle \cos\theta \rangle$:

$$\langle \cos\theta \rangle = \frac{\int_{-\pi}^{+\pi} \exp\left(-\frac{\Delta E(\theta)}{RT}\right) \cos\theta \, d\theta}{\int_{-\pi}^{+\pi} \exp\left(-\frac{\Delta E(\theta)}{RT}\right) d\theta} \tag{2.8}$$

Equation (2.8) is based on the integral of a Boltzmann term of the torsional-angle-dependent potential energy, $\Delta E(\theta)$, from Figure 14. With this value, we can append a term for the torsional angle dependence to our chain model:

$$\langle \vec{r}^2 \rangle = Nl^2 \frac{1 - \cos\varphi}{1 + \cos\varphi} \cdot \frac{< 1 + \cos\theta >}{< 1 - \cos\theta >} \tag{2.9}$$

We have now developed a model that takes into account both the universal physics of a polymer coil, which is expressed by its universal scaling according to $\langle \vec{r}^2 \rangle^{1/2} \sim N^{1/2}$, and that also contains terms that account for the chemical specificity of a given type of monomer or structural unit, characterized by its bonding angle and its average torsional angle. These specific values are constant for each kind of repeating unit, which is why they are often summed up into polymer-specific parameters. The torsional angle dependence $\langle 1 + \cos\theta \rangle / \langle 1 - \cos\theta \rangle$ is often called the *obstruction parameter*, σ^2. It is determined by the bulkiness of the side groups of a repeating unit: the more energy is needed for these units to change their conformation, the higher this value. More often, the **characteristic ratio,** C_∞, is used; it includes all chemistry-specific parameters in the form of

$$C_\infty = \frac{1 - \cos\varphi}{1 + \cos\varphi} \cdot \frac{\langle 1 + \cos\theta \rangle}{\langle 1 - \cos\theta \rangle} \tag{2.10}$$

With that, we get the **general scaling law for ideal polymer chains:**

$$\left\langle \vec{\boldsymbol{r}}^2 \right\rangle = \boldsymbol{C_\infty N l^2}$$

The characteristic ratio is large for stiff polymer chains with bulky or like-charged substituents along the chain backbone, such as PS, poly(methylmethacrylate), or poly(sodium acrylate), whereas it is low for flexible chains with uncharged and unbulky substituents, such as PE. Note that for short chains, the characteristic ratio shows a dependence on the number of repeating units, N, as depicted in Figure 17; it is then abbreviated by the symbol C_N. From about 80–100 repeating units on, however, it reaches an N-independent plateau, as also shown in Figure 17; it is then abbreviated by the symbol C_∞.

2.2.4 The Kuhn model

When we recapitulate what we have done in the preceding sections, we see that we have taken up the most simplistic model and modified it with further terms to achieve

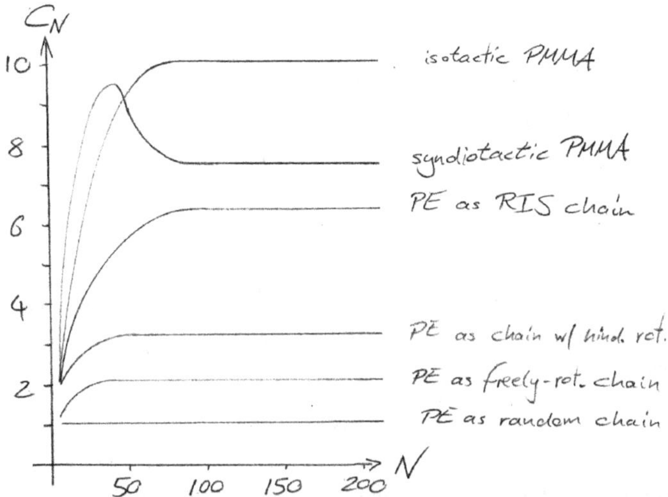

Figure 17: Characteristic ratio, C_N, as a function of the number of repeating units, N. Initially, C_N rises with N, but after about 80–100 repeating units, it reaches a plateau for each polymer. That is why the characteristic ratio is often referred to as C_∞. Note how its value raises from about 3 to about 10 if we go from flexible chains with unbulky substituents such as polyethylene to chains with bulky substituents such as poly(methylmethacrylate). Picture redrawn from H. G. Elias: *Makromoleküle, Bd. 2: Physikalische Strukturen und Eigenschaften* (6. Ed.), Wiley VCH, **2001**.

a closer resemblance to reality, thereby making it more complex. But wouldn't it be magnificent to have a model at hand that would retain the original simplicity of the random chain and, at the same time, still capture chemical specificity? This can be done on the basis of an ingenious approach by Hans Kuhn, which is the Kuhn model. In this model, a number (not necessarily a natural number) of several segments of a polymer chain with length l are grouped such to create a new conceptual polymer chain made up of new conceptual segments, the so-called Kuhn segments of new conceptual length l_K, the so-called Kuhn length (see Figure 18). The original chain end-to-end distance is retained for this new conceptual chain. The ingenious element of this approach is the following: the chemical specificity of the original polymer is incorporated into the model by normalizing the number N and length l of the repeating units to the characteristic ratio C_∞. That way, each given polymer "forgets its chemical specificity" and can be assessed by a **universal power law:**

$$\langle \vec{r}^2 \rangle = C_\infty N l^2 = \frac{N}{C_\infty}(C_\infty l)^2 = N_K \cdot l_K^2 \tag{2.11}$$

The new conceptual chain is made from $N_K = N/C_\infty$ Kuhn segments of length $l_K = C_\infty l$. The bonding angle φ between these segments is now random, whereas it was a fixed constant in the original chain (Figure 18). As a result, the Kuhn chain can be assessed like a random chain. With that approach, we have traced our more

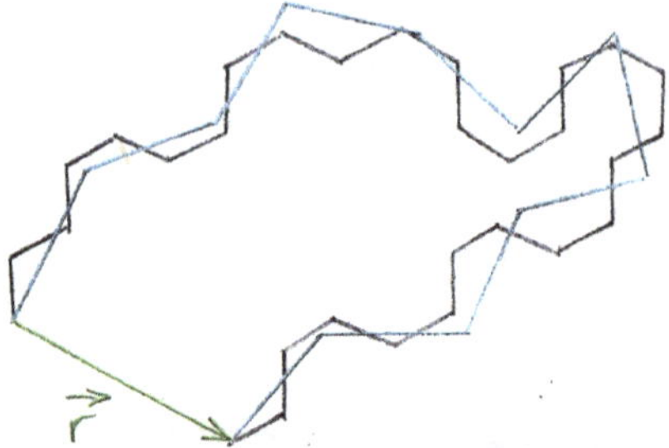

Figure 18: The *Kuhn model* sums up several actual polymer chain segments of length l to create a new conceptual polymer chain made up of N/C_∞ Kuhn segments of conceptual length $l_K = C_\infty l$, the Kuhn length. The end-to-end distance of the original chain, $\vec{r}$, is retained. The bonding angles of the Kuhn segments, however, are now random, whereas the bonding angles of the original chain were fixed. As a consequence, the conceptual chain can be assessed as a random chain.

and more complex models described in the previous sections back to their simple origin. Note that with $R^2 = C_\infty N l^2$ as well as $l_K = C_\infty l$ we get an interesting formula: $R^2 = C_\infty N l^2 = C_\infty \, lNl = l_K l_{cont}$, with $l_{cont} = Nl$ the contour length, that is, the chain length in an hypothetic fully decoiled conformation.

Hans Kuhn (Figure 19) was born on December 5, 1919, in Bern in Switzerland. He studied chemistry at the ETH Zürich and obtained his PhD degree at the University of Basel. He worked together with Linus Pauling at California Institute of Technology and Niels Bohr at the University of Copenhagen as a postdoctoral researcher, before he held professorships in Basel, Marburg, and the Max Planck Institute for Biophysical Chemistry in Göttingen. He died on November 25, 2012, in Basel at the age of 92.

Figure 19: Portrait of Hans Kuhn. Image from Wikimedia Commons, originally uploaded from private property by CarsiEi on March 31, 2009, with general permission for reproduction.

2.2.5 The rotational isomeric states model

We can also do the opposite of the Kuhn model's simplicity and try to find a maximally realistic expression for C_∞ for a given specific chain; this is the approach of the rotational isomeric states (RIS) model.

It explicitly considers the energies of the most relevant conformational states T, G$^+$, and G$^-$, denoted as $\Delta E(\theta)$ in Figure 14, and then uses the Boltzmann distribution to quantitatively appraise their populations:

$$\frac{n_{\text{gauche}}}{n_{\text{trans}}} = 2 \cdot \exp\left(\frac{-\Delta E(\theta)}{RT}\right) \tag{2.12}$$

This equation determines the exact ratio of *gauche* and *trans* conformations in a given polymer, as shown in Figure 20. The factor of 2 before the exponential term is due to the existence of two *gauche* conformations, G$^+$ and G$^-$. In addition to the *gauche*-to-*trans* conformational ratio, the RIS model also appraises how the conformation of each bond i is depending on what the conformation of the preceding bond $i-1$ is; in other words, the model quantifies how many dyads of TT, TG, and GG are present in the chain. This is done by quantifying the energies of these different combinations and translating them into relative population ratios, as summarized in Table 5.

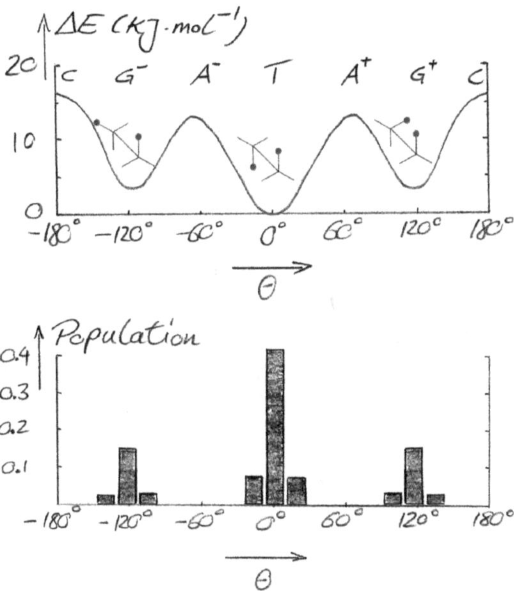

Figure 20: The RIS model quantitatively appraises the populations of the low-energy conformational states T, G$^+$, and G$^-$, and it determines the population of each of these states with the Boltzmann distribution. The upper diagram is a replot of Figure 14 and shows the energies of monomer–monomer bond torsional angles for butane; the lower diagram displays the relative populations of these bond angles at a temperature of 300 K. Upper diagram redrawn from H. G. Elias: *Makromoleküle, Bd. 1 – Chemische Struktur und Synthesen* (6. Ed.), Wiley VCH, **1999**; lower diagram redrawn from R. H. Boyd, P. J. Philips: *The Science of Polymer Molecules*, Cambridge Solid State Series, **1993**.

Table 5: Relative population numbers of T and G conformations of a bond *i* in a polymer chain depending on the conformation T or G of the preceding bond *i* – 1.

Bond *i* → 0° (T)	+120° (G^+)	−120° (G^-)	Bond *i* – 1 ↓
1	0.54	0.54	0° (T)
1	0.54	0.05	+120° (G^+)
1	0.05	0.54	−120° (G^-)

2.2.6 Energetic versus entropic influence on the shape of a polymer

A key part of eq. (2.12) is the ratio of $\Delta E/RT$. If this ratio is high, then the chain primarily adopts the energetically most favorable shapes. In synthetic polymers, these are either all-*trans* or alternating TGTGTG or TTGGTTGG conformational shapes, depending on the bulkiness of the substituents and their optimal low-energy arrangement in space. The most well-ordered polymers in nature are proteins; they have very specific energetically highly favorable folded shapes, and it is this specificity that gives rise to their specific functions, because these specific outstanding ordered shapes make possible the lock-and-key principle that enables specific enzyme–substrate interactions. By contrast, at a low ratio of $\Delta E/RT$, a polymer chain adopts the entropically most favorable conformation, which is that of a random coil.

A related ratio is that of the energy of activation for torsional *change*, $\Delta E^{\ddagger}{}_{TG}/RT$. This ratio corresponds to the height of the energy *barrier* between the T and G states and thereby describes the dynamics of a polymer chain. A high ratio of $\Delta E^{\ddagger}{}_{TG}/RT$ is encountered for stiff polymer chains that carry bulky side groups, such as PS. A low ratio, on the contrary, is encountered for flexible polymer chains that carry unbulky side groups, such as PE. This ratio is directly reflected in one of polymers' most important property: the glass transition temperature, which is the temperature needed for activation of main-chain dynamics. This temperature is high for PS, whereas it is low for PE due to the less easy activation of bond conformational changes in PS than in PE. (See: again, here we have a structure–property relation!)

Both the latter relations are dependent on temperature, which is because

$$\Delta G = \Delta H - T\Delta S \tag{2.13}$$

In general, the change of Gibb's free energy, ΔG, needs to be negative for any process to proceed spontaneously. We can see on the right-hand side of eq. (2.13) that the change of enthalpy, ΔH, and that of entropy, ΔS, are connected to each other by temperature, T. It directly follows that at low temperature, the entropy term has a low significance, so that the enthalpy term dominates. As a result, a polymer coil expands and adopts a more ordered macroconformation at low temperature. This can even lead to higher-order chain assemblies such as polymer crystals. On the opposite, at high temperature, the entropy term dominates; here, a polymer coils up to a disordered macroconformation.

2.2.7 The persistence length

Due to the fixed bonding angles and somewhat preferred conformational arrangements of the bonds in a polymer chain, as captured by the latter three models above, each segment i imparts a directional preference onto the next following ones $i + 1$, $i + 2$, $i + 3$, ... along the chain. It therefore takes a certain number of following segments until the directional "memory" of a given first segment is "forgotten". This "directional memory" or **persistence** is captured in the form of a quantity named *persistence length.* Mathematically, it can be defined as the projection of all following bond vectors $i + j$ onto the direction of a given first one i, as shown in Figure 21.

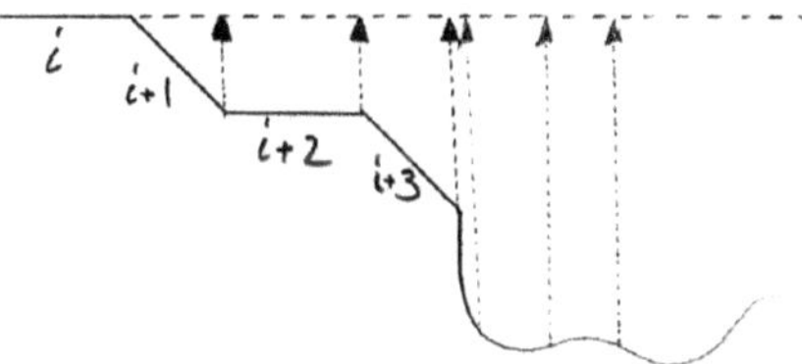

Figure 21: Projection of all bonds $i + j$ following a given one i onto its direction.

This projection is expressed as

$$l_p = \frac{1}{l_i} \sum_{j>i} \langle l_i \, l_j \rangle \tag{2.14}$$

Just as the characteristic ratio, C_∞, and the Kuhn length, $l_K = C_\infty l$, the persistence length captures the stiffness of a polymer chain. The stiffer the chain is, the longer is the length up to which the directional influence of a given segment is still recognizable. It is not surprising that the two quantities l_p and C_∞ are related to one another:

$$C_\infty = 2\frac{l_p}{l} - 1 \tag{2.15}$$

From that, we also obtain a relation between the persistence length, l_p, and the Kuhn length, l_K:

$$C_\infty + 1 \approx C_\infty = 2\frac{l_p}{l}$$

$$\Leftrightarrow C_\infty l = l_K = 2 l_p \tag{2.16}$$

As an example, consider PS. In this polymer, we have $C_\infty = 10.2$ and $l = 0.154$ nm; hence, based on eq. (2.15), we get $l_P = 0.86$ nm. This is still low! For comparison, consider double-stranded DNA, which has a persistence length of 63 nm. This extremely large value originates from the constrained primary constitution of the DNA double strand and from the multiple repulsive electrostatic interactions between the charges of its phosphate units. Chains with such high stiffness are no longer named flexible, but instead *semiflexible* or *wormlike*. In these, a different geometrical view on the persistence length (that reflects the same mathematics, of course) is commonly used, given by the intersection angle of two tangents on the chain, as shown in Figure 22.

Figure 22: Schematic of a semiflexible, wormlike chain with two tangents constructed to it that cross in a certain intersection angle.

In wormlike chains, the persistence length is the length that it takes to have the cosine of the angle shown in Figure 22 to decay down to $1/e$ (=0.368).

2.2.8 Summary

We have learned throughout this chapter that the most probable shape of polymer chain is that of a *random coil*. We have developed the simplest model to describe this coil, which is the *random chain model*. Based on that, we have continuously expanded this model to incorporate the chemical specificity of the monomer segments by adding contributions from the bonding angle, φ, in the *freely rotating chain model*, and the bond torsion angle, θ, for the *chain with hindered rotation* (see Table 6). This chemical specificity can be captured by the *characteristic ratio* C_∞, which is taken in the Kuhn model to resimplify and universalize the chain model while simultaneously retaining the given specificity of each given polymer. Furthermore, we have seen that the macroconformation of the chain is dependent on temperature and on the specific chemical design of the monomeric units.

Table 6: Summary of the characteristic simple chain models addressed in this chapter.

$R^2 = \langle \vec{r}^2 \rangle = C_\infty N l^2$	Random chain	Freely rotating chain	Chain with hindered rotation	RIS model
Bond length (l)	Fixed	Fixed	Fixed	Fixed
Bond angle (φ)	Free	Fixed	Fixed	Fixed
Torsion angle (θ)	Free	Free	Fixed average	Discrete: T, G^+, G^-
Char. ratio (C_∞)	1	$\frac{1-\cos\varphi}{1+\cos\varphi}$	$\frac{1-\cos\varphi}{1+\cos\varphi}\cdot\frac{\langle 1+\cos\theta\rangle}{\langle 1-\cos\theta\rangle}$	Specific

Polymers with bulky side groups exhibit stiff chains at ambient temperatures, whereas polymers with small side groups are flexible at ambient temperatures. We have therefore created our first rational structure–property relation.

To get an impression on some numbers, consider PE with $N = 20{,}000$, $l = 0.154$ nm, and $C_\infty = 6.87$. Whereas the length of the maximally elongated chain, $r = Nl$, would be 3080 nm, its rms end-to-end distance is $R = \langle r^2 \rangle^{1/2} = (C_\infty N l^2)^{1/2} =$ 57 nm. This estimate clarifies two insights. First, the polymer is heavily coiled in its natural state. Second, in that state, it has a size that falls right into the colloidal domain.

2.3 The Gaussian coil

LESSON 3: GAUSSIAN COILS AND BOLTZMANN SPRINGS

The last lesson has taught you that an ideal chain has the shape of a random coil. The following lesson will further refine this insight and make you understand how exactly the segmental density is distributed inside the coil. It will also show how the most fundamental thermodynamic property of such a coil, its entropy, reacts on deformation of the coil, thereby introducing the most relevant property of flexible polymers: their entropy-based elasticity.

We now want to take a closer look at how exactly a polymer coil looks like. We start by tying into the number example from the end of the preceding section and again consider PE with N =20,000, l = 0.154 nm, and C_∞ = 6.87, for which we have estimated an rms end-to-end distance of $R = \langle r^2\rangle^{1/2} = (C_\infty N l^2)^{1/2}$ = 57 nm. We now calculate the volume of just all the segments together, that is, the volume that a dense globule would have into which the coil might collapse: $V_{segments}$ = no. of segments × vol. per segment ≈ $N l^3$. For comparison, we also consider the volume that the coil has in its natural state: $V_{coil} = 4/3\ \pi R^3 = 4/3\ \pi (C_\infty N l^2)^{3/2} \approx 4\ C_\infty^{3/2} N^{3/2} l^3$. When we compare these two volumes by calculating their ratio, $V_{coil}/V_{segments} = (4\ C_\infty^{3/2} N^{3/2} l^3)/(N l^3) = 4\ C_\infty^{3/2} N^{1/2}$, we realize that this ratio is more than 10,000! This means that most of the volume of the polymer coil is actually *empty* and not occupied by the polymer chain material itself. But how, then, is that chain material distributed within the coil?

The polymer chain segments in a coil are distributed according to a **Gaussian radial density profile**:

$$c_{seg} = N\left(\frac{3}{2\pi R_g^2}\right)^{3/2} \cdot \exp\left(\frac{-3r^2}{2R_g^2}\right) \tag{2.17}$$

From the graphical representation of this equation, in Figure 23, we can see that the highest polymer segmental density is at the coil's center. For the blue curve in Figure 23, denoting a polymer with N = 20,000 segments, we have a segmental density of about 11 segments per nm^3 in the coil center; this corresponds to a molar concentration of about 20 mol L^{-1}. The further we move away from the coil center, the more does the density drop, which is especially true for short polymer chains with only a few segments, whereas longer chains have a less steep radial segmental density profile. Nevertheless, note that the degree of coiling, Q, gets more pronounced at high N, as shown in the following estimate:

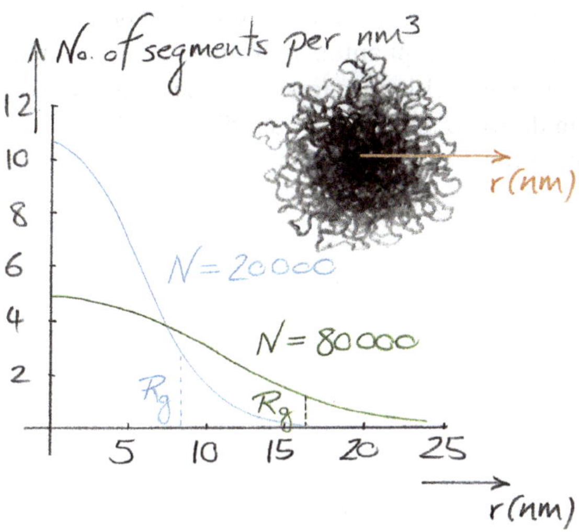

Figure 23: Concentration of polymer segments in a coil as a function of the coil radial coordinate, *r*, for two polyethylene chains with *N* = 20,000 and *N* = 80,000 repeating units. In both cases, the majority of segments is located at the coil center, which is more pronounced for the shorter of the two chains. In radial direction away from that center, the segmental density drops according to a Gaussian profile. The dashed lines indicate the radii of gyration of both coils. The upper-right sketch illustrates the radial segmental density profile in a schematic fashion.

$$Q = \frac{L_{\text{cont}}}{\langle r^2 \rangle^{1/2}} = \frac{Nl}{N^{1/2}l} = N^{1/2} \tag{2.18}$$

Hence, longer chains are coiled more notably than shorter chains, but in terms of their segmental density profile, long chains are more "dilute" at the coil center than short chains but have a higher segmental density at the rim in turn.

This last consideration is a good indication that we have now moved beyond average values to learn about the shape of a polymer coil. In the following, we will therefore look more closely at the coil statistics, with the aim to not only describe its *average* size, but the full *distribution* of it.

2.4 The distribution of end-to-end distances

2.4.1 Random-walk statistics

We have learned in Section 2.2 about the fundamental scaling law between the mean-square end-to-end distance and the number of monomer segments of an ideal polymer chain, $\langle r^2 \rangle \sim N$. We have also noted the similarity to the Einstein–Smoluchowski law of diffusion, $\langle x^2 \rangle \sim t$, and realized that this is due to very similar assumptions made in the two models: the path of a diffusing particle has no memory and can cross itself, and an ideal phantom polymer chain is assumed to be composed of volume-less segments without interactions, so it can cross itself, too. We now rely on this similarity to appraise the distribution of the chain end-to-end distance, r, by adopting **random-walk statistics.** We start by considering a one-dimensional random walk composed of N steps on the x-axis:

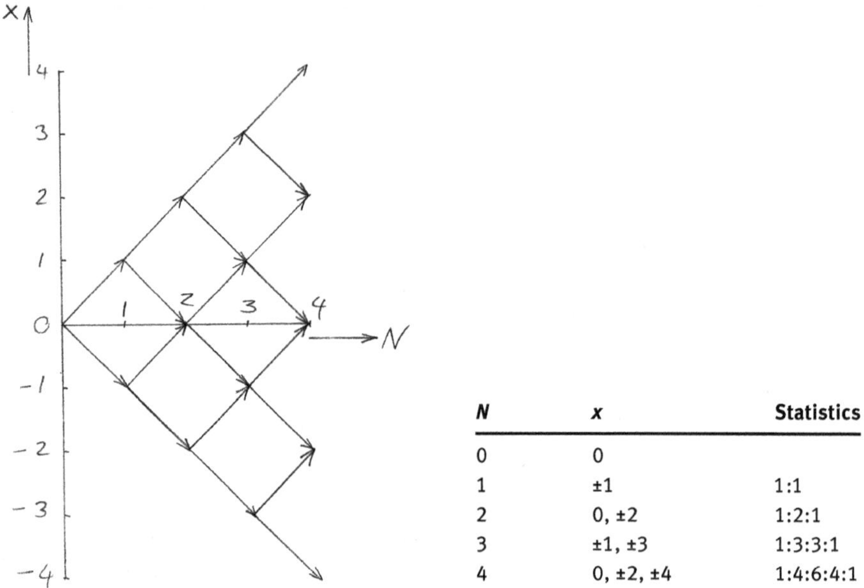

N	x	Statistics
0	0	
1	±1	1:1
2	0, ±2	1:2:1
3	±1, ±3	1:3:3:1
4	0, ±2, ±4	1:4:6:4:1

You may imagine this walk as follows: think that you place a marker at zero and flip a coin. Every time the coin lands on heads, you move the marker up in the x-direction, and every time the coin lands on tails, you move the marker down. After one flip, the marker can be at $x = \pm 1$ with a probability of 50% each. After two flips, the marker can either be back at $x = 0$, or it can be at $x = \pm 2$, with a probability of 50% for $x = 0$ and 25% for each $x = \pm 2$. In general, for a total number of steps $N = N_+ + N_-$, an end position of $x = N_+ - N_-$ can be calculated. It is apparent from the statistics column in the table above, which resembles Pascal's triangle, that the most probable outcome of any such random walk will be $x = 0$. This is because there's many walks that we can take for a given number of steps, but most of them lead us back to zero, whereas just few of them lead us far apart from zero. Using the binomial statistics behind the Pascal triangle in the table above, we can appraise the number of walks reaching position x after N steps as

$$W(N,x)=\frac{N!}{\left(\frac{N+x}{2}\right)!\left(\frac{N-x}{2}\right)!} \tag{2.19}$$

Equation (2.19) can be understood by illustrating the sequence of N steps as an ordered row of lots that we randomly draw out of a container. The first one that we draw is one out of a total of N, the second is one out of a remaining total of $(N-1)$, the third is one of a remaining total of $(N-2)$, and so forth. This gives us $N\cdot(N-1)\cdot(N-2)\cdot\ldots=N!$ possible permutations to arrange that sequence. Each lot has two possible values in a random walk: it can denote a step in the positive direction or a step in the negative direction. As a result, we can divide this sequence into two sub-ensembles. One comprises all those steps that go into the positive direction; we have a total of N_+ such steps. The other comprises all those steps that go into the negative direction; we have a total of N_- such steps. The order of these N_+ and N_- steps is irrelevant, because we always end up at the same final position of $x=N_+-N_-$. Hence, we have to divide our $N!$ by the number of permutations of these N_+ and N_- steps, which is $N_+!$ and $N_-!$ This gives us $N!/(N_+!\cdot N_-!)$ possible walks with a given number of positive and negative steps. We have seen already that the majority of these walks will lead us close to the center at around $x=0$, so we are particularly interested about those walks with $N_+=N/2+(x/2)$ and $N_-=N/2-(x/2)$. Replacing the N_+ and N_- in the expression $N!/(N_+!\cdot N_-!)$ with these terms turns it into eq. (2.19).

The total number of walks that we can take at all is 2^N. Together with eq. (2.19), this gives us the probability of reaching position x after N steps:

$$\frac{W(N,x)}{2^N}=\frac{1}{2^N}\cdot\frac{N!}{\left(\frac{N+x}{2}\right)!\left(\frac{N-x}{2}\right)!} \tag{2.20}$$

As $W(N,x)$ is a big number, we take the logarithm of it and apply Stirling's approximation $\ln(N!)\approx N(\ln(N)-1)$ to eliminate the factorials. We may then expand it into a Taylor series with x/N as the variable. This will give us a zeroth-order element that is independent of x/N, whereas we will have no first-order element, as the distribution is symmetrical. The second-order element will be of kind $(x/N)^2$. If we interrupt the series after this element, it will be of type $\ln[W(N,x)]\approx a-b\cdot(x/N)^2$, so we have $W(N,x)\approx\exp[a-b\cdot(x/N)^2]=\exp[a]\cdot\exp[-b\cdot(x/N)^2]$. The exact calculation yields

$$\frac{W(N,x)}{2^N}\cong\sqrt{\frac{2}{\pi N}}\cdot\exp\left(\frac{-x^2}{2N}\right) \tag{2.21a}$$

If we normalize our coordinate system from a step width of 1, as we had it in the sketch above, to a step width of ½, we transform the integer distance on the x-axis in the walks in the table above from 2 to 1. With that, we get the probability distribution of the displacement x in a one-dimensional random walk:

$$p_{1d}(N,x) = \sqrt{\frac{1}{2\pi N}} \cdot \exp\left(\frac{-x^2}{2N}\right) \tag{2.21b}$$

Equation (2.21b) differs from eq. (2.21a) just by a factor of 2, simply because the integer distance on the x-axis was 2 for the walks in the table above, whereas it has now been normalized to 1.

So far in this book, we have considered the mean-square end-to-end distance, which corresponds to the mean square displacement in the random walk. In general, such mean values can be calculated from distribution functions by the use of mathematical operators. Applying such an operator to eq. (2.21b) gives us the mean square of the distribution:

$$\langle x^2 \rangle = \int_{-\infty}^{+\infty} x^2 p_{1d}(N,x)\mathrm{d}x = \sqrt{\frac{1}{2\pi N}} \int_{-\infty}^{+\infty} x^2 \exp\left(\frac{-x^2}{2N}\right)\mathrm{d}x = N \tag{2.22}$$

This simple result provides a useful identity, as it connects N to $<x^2>$. With that, we can replace one by the other in eq. (2.21b) and obtain a formula with just one variable (x):

$$p_{1d}(x) = \sqrt{\frac{1}{2\pi <x^2>}} \cdot \exp\left(\frac{-x^2}{2<x^2>}\right) \tag{2.23}$$

We now want to move into the three-dimensional realm, so we need to consider all three spatial directions. Fortunately, a simple random walk does not have a preferred direction; in other words: we consider an isotropic case. In such a case, we may construct a three-dimensional random walk simply by the superposition of three independent components, one for each spatial dimension:

$$P_{3d}(\vec{r})\mathrm{d}r_x\mathrm{d}r_y\mathrm{d}r_z = p_{1d}(r_x)\mathrm{d}r_x \cdot p_{1d}(r_y)\mathrm{d}r_y \cdot p_{1d}(r_z)\mathrm{d}r_z \tag{2.24}$$

In this simple treatment, the three-dimensional end-to-end distance is additive of its three components:

$$\langle r^2 \rangle = Nl^2 = \langle r_x \rangle^2 + \langle r_y \rangle^2 + \langle r_z \rangle^2 \cdot \langle r_x \rangle^2 = \langle r_y \rangle^2 = \langle r_z \rangle^2 = \frac{Nl^2}{3} \tag{2.25}$$

With that, we can formulate an expression for a three-dimensional random walk as follows:

$$p_{3d}(\vec{r}) = \frac{n(\vec{r})}{\sum_r n(\vec{r})} = \left(\frac{3}{2\pi <r^2>}\right)^{3/2} \cdot \exp\left(\frac{-3r^2}{2<r^2>}\right) \tag{2.26}$$

In this equation, $n(\vec{r})$ is the number of walks (or polymer chains) with a given displacement (or chain length) $\vec{r}$, and $\sum_r n(\vec{r})$ is the total number of walks (or chains).

Let us now examine how a graphical representation of eq. (2.26) looks like. As an example, a three-dimensional random-walk-type probability distribution of the end-to-end distance for the example of a PE chain with N = 20,000 repeating units is shown by the green curve in Figure 24. It visualizes the probability to have an end-to-end vector $\vec{r}$ with a given length $|\vec{r}|$ and a given direction. This corresponds to the probability of finding the second chain end in a certain specific volume element at a distance and direction $\vec{r}$ from the origin if the first chain end is placed right there. This probability is maximal at an end-to-end vector of zero, as the random-walk statistics has given us the highest likelihood of obtaining that very displacement,

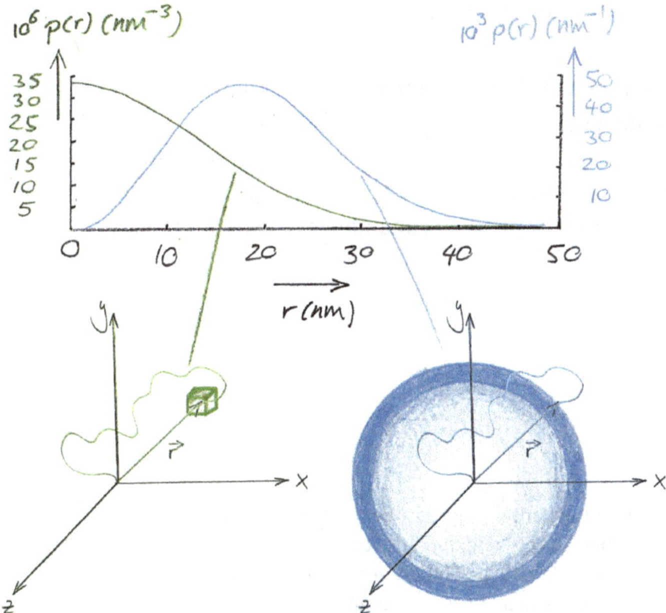

Figure 24: Distribution of end-to-end distances of an ideal polyethylene coil composed of N = 20,000 repeating units. The green curve is the outcome of a three-dimensional random-walk statistical treatment (eq. (2.26)). Here, the most likely trajectory, or end-to-end vector, $\vec{r}$, is zero. This corresponds to the likelihood of finding the second chain end exactly in a specific small volume element in space if the first chain end is placed in the origin, as illustrated in the lower left sketch. The blue curve filters off all directional dependences by multiplication with a sphere surface area (eq. (2.27)), thereby reflecting the distribution of end-to-end vector *lengths* without caring about into which direction they point. This corresponds to the likelihood of finding the second chain end localized anywhere on a spherical orbit with radius $|\vec{r}|$ around the first end that is again placed in the origin, as illustrated in the lower right sketch. (Note that the green $p(r)$ is actually of type $p(\vec{r})$, whereas the blue $p(r)$ is actually of type $p(|\vec{r}|)$. For common representation on the same simple r-axis, they are both denoted in a simplified form as $p(r)$ on both ordinates. Their mathematical difference is apparent from the different physical units, though).

whereas the probability distribution for other end-to-end vectors or displacements drops in the form of a Gaussian bell curve.[15]

In elementary statistics, the likelihood of an event is proportional to the frequency of its occurrence in a large ensemble. Hence, the Gaussian probability distribution that we have just found is directly related to the Gaussian distribution of the monomer segmental density from Section 2.3, which is also highest at the center of the polymer coil. Due to that identity, ideal polymer chains are often named **Gaussian chains.**

We can filter off the directional dependence of the random walk by multiplying it with the surface area of a sphere of radius r:

$$p_{3d}(|\vec{r}|) = 4\pi r^2 \left(\frac{3}{2\pi\langle r^2\rangle}\right)^{3/2} \cdot \exp\left(\frac{-3r^2}{2\langle r^2\rangle}\right) \tag{2.27}$$

That way, we generate a function that expresses the probability distribution of end-to-end vector lengths, independent of their direction.[16,17] The blue curve in Figure 24 reflects this relation for the same PE compound considered above. It visualizes the probability to find *any* end-to-end vector of length $|\vec{r}|$ independent of its direction, that is, any end-to-end vector on a spherical circumference with radius r.

15 Equation (2.26) is similar to an equation that you (hopefully) know from elementary physical chemistry: the velocity distribution in the kinetic theory of gases, $p(\vec{v})$. That distribution has the form of a Gaussian bell curve as well, as it essentially reflects the Boltzmann distribution of the (kinetic) energies of the gas particles, which has its maximum at zero.

16 Again, something similar is known to you from the kinetic theory of gases, where $p(\vec{v})$ turns into $p(|\vec{v}|)$, the Maxwell–Boltzmann distribution, by the same mathematical operation.

17 Even more, you may have encountered a similar concept earlier in yet a different field of elementary physical chemistry already. In quantum mechanics, the square of the wave function of the s-orbital is considered as a measure of the likelihood of finding the electron in a certain point in space. This likelihood is maximal at a distance of zero from the nucleus and then drops when moving to larger distances. Just as with the ideal-polymer case treated here, in this problem, though, the directional dependence is not of so much interest than the radial-distance-dependence. To account for that, the square of the wave function is multiplied with a spherical surface area of radius r to obtain a function that gives the likelihood of finding the electron at a certain distance from the nucleus, independent of the direction. Whereas the first function (the square of the wave function) drops with the radial coordinate, the second function (the sphere surface-area function) increases as a power law (with a power of 2). The product of both functions first raises, then reaches a maximum, and then drops with the radial coordinate, with the intermediate maximum denoting the atomic radius.

2.5 The free energy of ideal chains

We now have a formula at hand that reflects how many microconformations, characterized as three-dimensional random walks, lead to a certain chain macroconformation, characterized by its mean-square end-to-end distance. This is information about the structure, about the shape of the polymer chain. In this book, however, we have set out to build structure–property relations. So, is it possible to use the structural information we have obtained and extract from them information about the polymer's properties? Would it be possible, for example, to calculate thermodynamic quantities from them? As it turns out, this is feasible: we can use statistical thermodynamics, which derives macroscopic thermodynamic quantities from the likelihood of microscopic states of a system. In statistical thermodynamics, Boltzmann's entropy formula connects the entropy, S, to the number of possible microconfigurations, W, as $S = k_B \ln(W)$. When we insert the probability distribution of the end-to-end distances into that formula, we can derive an expression for the entropy, S, of an ideal chain:

$$S = S_0 + k_B \ln W(N, \vec{r}) \cong S_0 - \frac{3k_B r^2}{2\langle r^2 \rangle} \tag{2.28}$$

where S_0 is the entropy at an end-to-end distance of zero. We see that the total entropy, S, is maximal (with a value of S_0) for an end-to-end distance of $r = 0$, because the probability of obtaining that distance in a random walk is maximal (green curve in Figure 24).

From this, we can further formulate the free energy, F, of an ideal polymer chain:

$$F = U - TS = F_0 + \frac{3k_B T r^2}{2\langle r^2 \rangle} \tag{2.29}$$

where U is the internal energy and F_0 is the free energy at an end-to-end distance of zero. In the case of an ideal polymer chain, U is independent of the end-to-end distance, because we imagine the chain segments to have no interactions, so energy doesn't care about their spatial arrangement and microconformations. This means that any change of the ideal-chain free energy stems from entropy alone.

2.6 Deformation of ideal chains

2.6.1 Entropy elasticity

When we slightly stretch an ideal polymer chain, the most feasible microscopic process of accounting for that deformation is decoiling of the chain, that is, transformation of local *gauche* or *cis* conformations into *trans* conformations, as this does not cost any significant amount of energy (see Figure 14). With that, though, we reduce the number of possible microconformations that realize the coil's macroconformation, thereby forcing it into an entropically more unfavorable state. As a result, as soon as the deformation cedes, the chain will relax back to the most probable coiled structure. This phenomenon is based on entropy and is therefore called **entropy elasticity**.

When we differentiate the free energy with respect to r_x, we can calculate the force needed for a deformation in the x-direction:

$$\vec{f}_x = \frac{\partial F(N,\vec{r})}{\partial r_x} = \frac{3k_B T}{\langle r^2 \rangle} \cdot \vec{r}_x = \frac{3k_B T}{Nl^2} \cdot \vec{r}_x \tag{2.30}$$

The latter formula tells us how much force is needed to stretch a chain by a distance r_x. Its form is that of Hooke's law ($f = \kappa \cdot x$), which connects the force necessary to achieve a certain extent of deformation of a body, $f \sim x$, by its spring constant, κ, a fundamental material property expressing how good mechanical deformation energy can be stored and released, and hence, how much the material deforms upon application of a given force. In the case of an ideal polymer chain, we can consider $(3k_B T)/Nl^2$ to be an **entropic spring constant**. As this quantity only contains structural information, namely, the number of segments and the segmental length, we have established another structure–property relation.

Strictly speaking, eq. (2.30) and the concepts we have derived from it are only valid for Gaussian chains in terms of the Kuhn model, that is, chains with a free segmental bonding angle φ. How would this picture change when we would consider it for the freely rotating chain that has a fixed segmental bonding angle φ? This question can be answered on two levels. On a conceptual level, the freely rotating chain is already more ordered than a Gaussian chain. It therefore already has a lower entropy, such that a further loss of entropy would not have such a significant influence, which means that these chains should be easier to stretch. On a mathematical level, we would have to consider the characteristic ratio C_∞ as a further factor in the denominator of the entropic spring constant. This would result in a smaller entropic spring constant, which would also mean that freely rotating chains are easier to stretch.

The entropic spring constant $3k_B T/Nl^2$ has a temperature dependence in its numerator. According to this dependence, it will rise with increasing temperature, thereby making it harder to deform an ideal polymer chain at higher temperature. This effect can be understood because entropy in thermodynamics always occurs along with temperature in the form of $T\Delta S$. Hence, the reduction of entropy upon chain stretching is even more pronounced at higher temperature, making it more unfavorable and therefore harder to realize.

So far, with eq. (2.30), we operate at a single-chain level. When we consider an ensemble of n chains, the essence of eq. (2.30) still holds for each single chain, such that the total force needed to deform all n of them is just n times that of a single chain, so a factor n will enter the right side of eq. (2.30). Furthermore, if we normalize the force to an area, this translates into a pressure, p, on the left side of the equation, whereas the length l in the denominator on the right side merges with that newly introduced area to a volume. We then end up with an expression strikingly close to the ideal gas law: $p = nk_BT/V$. Once more, we have discovered astonishing similarity between the physics of the ideal polymer chain and the ideal gas. For our present discussion on entropy elasticity, this is straightforward to understand. According to the ideal gas law, the pressure of an ideal gas rises if we compress it. Why is that? Energetically, the point-like gas molecules do not care about their mutual distance, and with that, about the volume we give them, because they do not have interactions, neither attractive nor repulsive. Entropically, however, a reduction of the system's volume reduces the number of possibilities for the molecules to arrange themselves in space. In short: a reduction of the volume reduces the *freedom* of the gas molecules. This comes along with an entropic penalty, which translates into an increase in free energy, and thus, to a backdriving force just like the one according to eq. (2.30), which translates into a pressure if we normalize it to area. The same is seen by a rise of the stress, $\sigma = f/A$, in a polymer sample that is subject to stretching, which is also caused by the loss of conformational freedom, and hence, the entropy penalty that comes along with that.

In contrast to the entropic origin of elasticity of ideal polymer chains and ideal gases, the deformation of a classical solid such as a wire of metal is fundamentally different. Here, upon deformation, the metal atoms are lifted out of their equilibrium positions in the crystal lattice that correspond to a minimum in their (Lennard–Jones-type) interaction potential, and as a result, an energetic-based restoring force arises. Upon increase of temperature, such an energy-elastic body will expand, because the additional energy enables the atoms to oscillate more extensively around their energy-minimum positions, and due to the skew shape of the (Lennard–Jones-type) interaction potential well (which has a steeper incline at its left than at its right rim), this stronger wiggling corresponds to a shift of the average atom positions in the lattice to larger separations. This energetic excitation also allows the material to be deformed easier at higher temperatures. Ideal polymer chains, by contrast, will shrink at elevated temperatures, as an energetically favorable but entropically unfavorable excess of local *trans* conformations in the chains will get less dominant in this case, due to the fundamental coupling of temperature and entropy in the form of $T\Delta S$. As a result, the chain end-to-end distances in a rubber sample under slight load shrink upon heating in order to achieve their entropically most favorable value of zero, and yet in turn, the whole sample specimen shrinks. An ideal gas, yet in turn, does not contract when temperature is increased, but instead, it shows thermal expansion, because its volume is directly proportional to temperature ($pV = nRT$). All these various differences and similarities of the two fundamental examples of entropy-elastic materials, which are the ideal gas and ideal polymer chains, as well as the classical example of an energy-elastic material, which is a piece of metal wire, are illustrated in Figure 25.

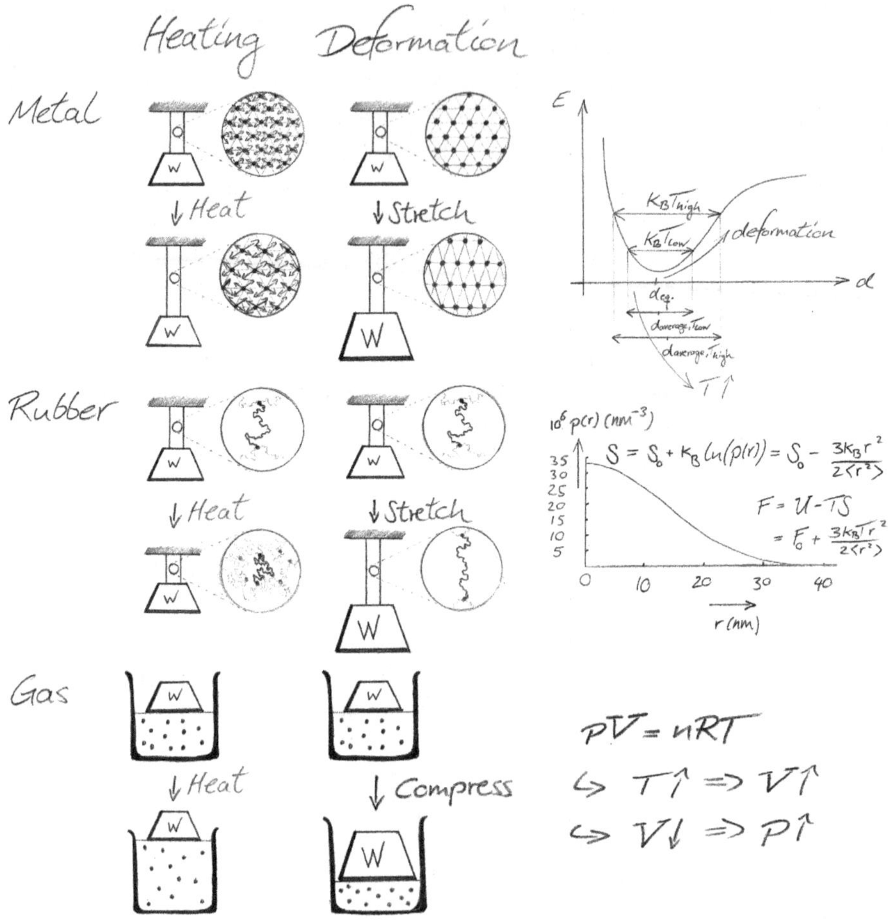

Figure 25: Overview of the different possible modes of deformation for classical solids (metal), polymers (rubber), and gases. Upon heating, a classical solid and a gas expand, whereas a polymer such as rubber shrinks. In a classical solid, this is because the atoms in a crystal lattice oscillate more heavily around their equilibrium positions at higher temperatures, and due to the skew shape of their Lennard–Jones-type interaction potential well (which has a steeper incline at its left than at its right rim), this stronger wiggling corresponds to a shift of the average atom positions in the lattice to larger separations. In a gas, even more simply, the basic equation of state $pV = nRT$ explains expansion on rise of temperature. Ideal polymer chains, by contrast, shrink at elevated temperatures, as an energetically favorable but entropically unfavorable excess of local *trans* conformations in the chain will get less dominant in this case, due to the fundamental coupling of temperature and entropy in the form of $T\Delta S$; as a result, the chain end-to-end distances in a rubber sample under slight load shrink upon heating in order to achieve their entropically most favorable value of zero, and yet in turn, the whole sample specimen shrinks. Upon mechanical deformation, the atoms of a classical solid are moved out of their potential energy minima, thereby creating a restoring force based on energy. Deformation of a polymer and a gas, by contrast, reduces the number of conformations or freedom of arrangement of the molecules in space, respectively, thereby creating a restoring force based on entropy. Picture inspired by J. E. Mark, B. Erman: *Rubberlike Elasticity: A Molecular Primer* (2nd ed.), Cambridge University Press, **2007**.

2.6.2 A scaling argument for the deformation of ideal chains

As an alternative to the lengthy statistical treatment of the entropic elasticity of ideal polymer chains that we have just discussed in the preceding sections, Rubinstein and Colby have introduced a clever treatment based on a **blob concept** (that originates from De Gennes). In this concept, the chain is regarded as a sequence of blobs of size ξ, each containing g segments. Up to the blob length scale, k_BT is the most relevant energy. As a consequence, any external energy is smaller than k_BT inside each blob, it is exactly k_BT at the blob scale ξ, and it is larger than k_BT above the blob scale. When the chain is stretched by an external force f_x, the segments inside the blobs are unaffected by this deformation, because on scales up to the blob scale ξ, the deformation energy is weaker than – or, in other words, *screened by* – the ever-present thermal energy k_BT. As a result, the deformation is effective only on larger length scales, where enough deformation energy is cumulated to outweigh k_BT. In this view, the chain segments inside the blobs always obey the ideal scaling law

$$\xi^2 = gl^2 \tag{2.31}$$

which is a blob-scale variant of the basic law $R^2 = \langle \vec{r}^2 \rangle = Nl^2$ that we have discussed in terms of the ideal chain in Section 2.2.1 (eq. (2.4)). In contrast to these ideal statistics on small scales, the stretched length of the entire chain can be viewed as a unidirectional linear sequence of blobs, as shown in Figure 26.

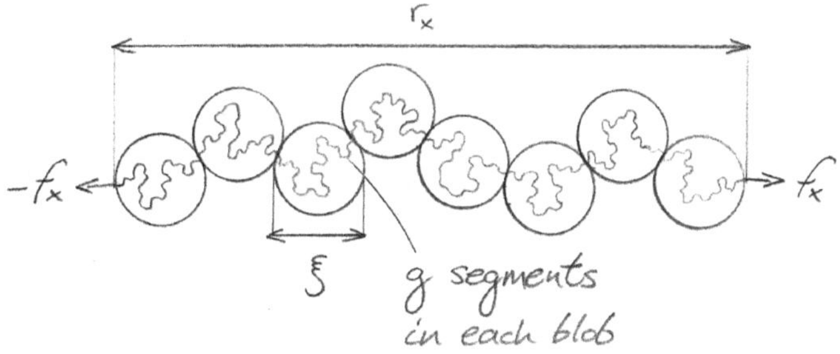

Figure 26: Schematic representation of the *blob concept*. Here, a chain with length r_x is viewed to be composed of conceptual units named *blobs* of a limiting length scale ξ, each containing g of the actual monomer segments. At scales smaller than the blob size, any external energy is smaller than k_BT. This means that when the chain is stretched by a force f_x, the deformation energy inside each blob is screened by k_BT, and deformation is only effective on length scales longer than ξ. Picture redrawn from M. Rubinstein, R. H. Colby: *Polymer Physics*, Oxford University Press, **2003**.

Mathematically, a unidirectional linear sequence of blobs is expressed as the number of blobs, (N/g), times the blob size ξ:

$$r_x = \frac{N}{g}\xi \tag{2.32a}$$

Replacing g in the denominator by eq. (2.31) transforms this into

$$r_x = \frac{Nl^2}{\xi} \tag{2.32b}$$

That can be rearranged to

$$\xi = \frac{Nl^2}{r_x} \tag{2.33}$$

Plugging that into eq. (2.32) gives

$$g = \frac{N^2 l^2}{{r_x}^2} \tag{2.34}$$

These latter formulae show that at stronger deformation (larger r_x), the blobs get smaller, meaning that the length scale up to which the deformation is screened by the thermal energy gets smaller. At maximal deformation ($r_x = Nl$), the blob size is $\xi = l$ (and $g = 1$), which means at such extreme deformation, the blobs have shrunken down to the size of the actual monomer; then, the deformation is notable on all length scales.

When the chain is conceived as such a sequence of blobs and then stretched in x-direction, the blob sequence gets ordered from being random to being aligned in that direction. This ordering comes along with a loss of one directional degree of freedom per blob, thereby raising the chain's free energy by that very extent. Hence, the free energy of stretching in the x-direction, F_x, is one $k_B T$ increment per blob:

$$F_x = k_B T \frac{N}{g} = k_B T \frac{{r_x}^2}{Nl^2} \tag{2.35}$$

This yields the force for deformation, f_x, in the limit of moderate stretching:

$$f_x = \frac{\partial F}{\partial r_x} \approx k_B T \frac{r_x}{Nl^2} \tag{2.36a}$$

This result is qualitatively similar to that from the longer exact derivation in the preceding section, but with Rubinstein's and Colby's blob concept and scaling approach, it was obtained much quicker and easier. This is a great advantage of scaling discussions as the one just led: they give results in good semiquantitative agreement to those obtained from exact and often more complicated derivations, but they do so in a much simpler and quicker fashion. Another great advantage is that they make us see conceptual grounds. We may get such an insight by replacing r_x/Nl^2 in the last equation by eq. (2.33), yielding

$$f_x = \frac{\partial F}{\partial r_x} \approx k_B T \frac{r_x}{Nl^2} \approx \frac{k_B T}{\xi} \tag{2.36b}$$

Thus, we can conceive that the deformation energy is $k_B T$ per blob. As shown above, the blobs get smaller as the deformation gets stronger (see eq. (2.33)). The greatest extent of deformation is if the chain is fully expanded to a rod-like object of length Nl. In that extreme, according to eq. (2.33), the blobs have shrunken down to the monomer segmental length of l (and accordingly, following eq. (2.34), the number of monomers per blob is then just 1). In that extreme, according to eq. (2.36b), the energy for deformation is $k_B T$ per monomer.

The reason why the change in scale that we have just used is valid and can still be described by the same mathematical equations is the fact that polymers are *fractal* and *self-similar* objects, a topic that we will discuss in the following paragraph.

2.7 Self-similarity and fractal nature of polymers

The relations between the mass and the characteristic size of any geometrical object can be described by scaling laws. A three-dimensional sphere, for example, exhibits scaling of $m \sim r^3$. A two-dimensional piece of paper exhibits scaling of $m \sim r^2$, and a one-dimensional piece of wire exhibits scaling of $m \sim r^1$. Generally, any object exhibits scaling according to

$$m \sim r^d \tag{2.37}$$

where d is its geometrical dimension.

The same principle is valid for ideal polymers; these obey the basic scaling law $R \sim N^{1/2}$ that we have developed in Section 2.2.1. N is proportional to their mass ($m = N \cdot m_{monomer}$); thus, comparison to the general relation $m \sim r^d$ denotes ideal polymers to have a dimension of 2. This is a so-called **fractal dimension**, as it is different from the geometrical dimension, which is 3 for a polymer in our three-dimensional world. In the form introduced here, the fractal dimension is that of a *mass fractal*, as it establishes a relation between the object's mass and its size. This fractality is not limited to ideal polymer chains. As we will see in the next main chapter, the general scaling law of a real polymer chain is

$$R \sim N^{\nu} \tag{2.38}$$

with $\nu = 1/d_{fractal}$, the *Flory exponent*.

Table 7 compiles a glimpse on fractal dimensions of various polymer types. In general, a smaller fractal dimension denotes the object that it belongs to be less dense. If we compare the fractal dimensions of ideal polymer chains, which is 2, to that of real chains with short-range repulsion, which is 5/3, we see that the latter is smaller. This means that real chains are less dense than ideal chains. The reason

Table 7: Overview of the fractality of ideal and real polymer chains with linear or branched architecture.

Architecture	Interactions	Spatial dimension	Fractal dimension
Linear	None	Any	2
Linear	Short-range repulsion	2	4/3
Linear	Short-range repulsion	3	5/3
Branched	None	Any	4
Branched	Short-range repulsion	2	8/5
Branched	Short-range repulsion	3	2

for that is the short-range repulsion between the monomer segments in a real chain, which pushes them apart from one another, thereby resulting in coil expansion that comes along with a lower segmental density inside the coil.

Fractal objects are also **self-similar**. This concept can be illustrated by the example shown in Figure 27(A). A two-dimensional square with side length L and area A has a scaling law for its area of $A = L^2$. A subsquare within that plane with a side length l and area a has the same basic shape, and it also exhibits an area scaling law of $a = l^2$. Without information on the actual length scale that we look at, we can therefore not distinguish whether we look at the whole object or a subunit of it. This phenomenon is called *self-similarity*. Many objects in nature are self-similar: clouds, coastlines, the surface of broccoli, and more. For all these objects, when

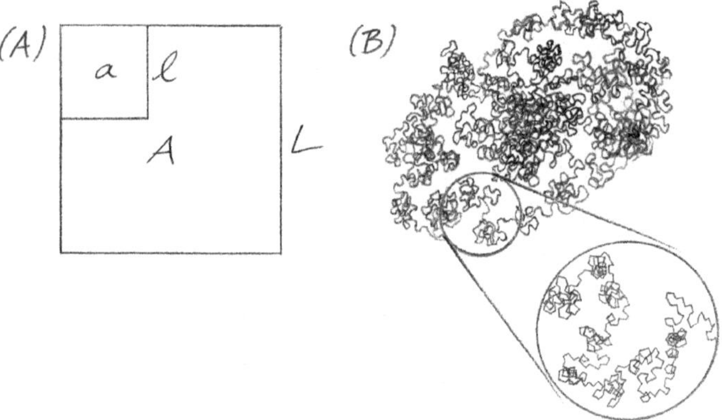

Figure 27: Illustration of the concept of *self-similarity*. **(A)** Consider a two-dimensional square with side length L and area A, and within that, a subsquare with side length l and area a. Both squares exhibit the same scaling law for their area, but on different length scales: $A = L^2$ and $a = l^2$. These laws are *self-similar*. **(B)** The same principle holds true for polymers: The basic scaling law of an ideal chain is $R \sim N^{1/2}$, which has a variant for subsegments of the chain of $\xi \sim g^{1/2}$. Again, these laws are *self-similar*. Picture in (B) is redrawn from M. Rubinstein, R. H. Colby: *Polymer Physics*, Oxford University Press, **2003**.

being shown a picture of them, you cannot tell whether you see a small or large section of the object, as it has the same appearance on different scales.

The shape of an ideal polymer coil is also self-similar, as shown in Figure 27(B). Although the shape of the whole chain shown in this figure is not *exactly* equal to that of the magnified chain subsection, *on average* the same random-walk-type sequence is seen on these different length scales. With that, also the scaling law of the ideal chain applies on both these (and further) different length scales. For the whole chain of N monomers, we have a scaling of $R \sim N^{1/2}$. For a subunit of the chain with just g monomers, we have a similar scaling of $\xi \sim g^{1/2}$. Rearranging these two scaling equations yields:

$$R = N^{1/2} l \Leftrightarrow l = R N^{-1/2} \tag{2.39a}$$

$$\xi = g^{1/2} l \Leftrightarrow l = \xi g^{-1/2} \tag{2.39b}$$

We can combine these latter two equations via their common variable l to obtain

$$R N^{-1/2} = \xi g^{-1/2} \Leftrightarrow R = \left(\frac{N}{g}\right)^{1/2} \xi \tag{2.40}$$

This equation has again the same form as the one we started with: it relates the polymer coil size, R, to the square root of a dimensionless number times an elementary unit size. In this notation, the chain can be viewed as a sequence of (N/g) segments of size ξ each, which was just the notation taken for the development of the blob concept in Section 2.6.1, where ξ was set to be the blob size.

Eq. (2.40) describes our chain, which is actually a random walk of N segments of size l each, as a new conceptual random walk of (N/g) blobs of size ξ each. We have done something similar before in Section 2.2.4: in the Kuhn concept, we have also renormalized our actual chain of N segments of size l each as a new conceptual chain of N_K segments of size l_K each. Now, the same was done with blobs as the new conceptual segments. Due to the self-similarity of polymer chains, this kind of renormalization works with any new scale, as long as we stay on scales larger than the Kuhn length l_K, because below that length, the polymer chain is no longer universal and self-similar but markedly exhibits its chemical specificity. In general, such kind of **scale transformation** works with any object that is self-similar and therefore **scale invariant**, meaning that it does not have a natural length scale that sets its further properties (like its mass or surface area). Scale invariance is a fundamental phenomenon in nature, similar to symmetry.

As a summary, in both the Kuhn model and in the blob concept, we renormalize a chain by a sequence of new conceptual rather than the actual repeating units and thereby shift the segmental length and number to new scales. The only mathematical function that allows such renormalization to be performed is *power laws*, which are also named *scaling laws* for that reason. If another kind of function, for example, a transcendental function[18] like $R = R_0 \ln(N/N_0)$, would describe the N-dependence of R, such rescaling would be mathematically impossible. Again, the reason for this is the absence of a natural scale (as it would be captured by N_0 and R_0 in the latter hypothetic equation) in objects that are scale invariant. Hence, the characteristics of self-similar and therefore scale-invariant objects such as polymers are always power-law type. This is the reason why power laws occur all over this textbook; they are inherent to polymers, as polymers are self-similar and fractal. [One further note: it is always *single* power-law terms that describe the relations between the characteristic quantities of self-similar objects (such as the N-dependence of R of a polymer), but not sums of them. Imagine, for example, that a sum of two power-law terms would describe the N-dependence of R in a form like $R = A\cdot(Nl) + B\cdot(Nl)^2$. Then, there would be coefficients A and B with different physical dimensions; A would be dimensionless, while B would have the unit m^{-1}. Their ratio A/B would then have a unit of m and therewith again reflect a natural length scale in the object under consideration.]

18 In a transcendental function, the argument is not allowed to have a physical unit; typical examples are exp, ln, sin, cos, and so on.

3 Real polymer chains

Lesson 4: Real Chains

In your elementary classes on physical chemistry, at some point, the ideal gas model reached its limitations, and it was necessary to expand it such to account for the finite volume as well as interactions of the gas particles. A similar point is reached when interactions and the own volume of polymer-chain segments can no longer be disregarded. This lesson accounts for both and thereby introduces an elementary quantity named *excluded volume*. This quantity, in turn, allows a spectrum of different types of solvents to be defined for polymers.

Now that we have examined and understood the ideal chain model, we can take the next step and model a polymer chain that resembles the real world more closely. This leads us to the **real chain model**. We approach it by readjusting our premises. So far, we have imagined a polymer chain to consist of monomer segments that have no volume and show no interactions, neither with one another nor with a solvent in the surrounding. Now, we overcome these simplifications and explicitly consider that the chain segments have a finite *covolume* and that they have *interactions*.[19] Our first focus is on appraising how these two effects affect the shape of the real polymer chain. We may get a first notion on that by pure intuition. If the monomer segments of a real chain have a finite volume, then two of them cannot occupy the same spot in space, as it was hypothetically possible for an ideal phantom chain with a self-crossing random-walk-type coil shape, as shown in the left sketch of Figure 28. A real chain, by contrast, has the shape of a **self-avoiding random walk**, as shown in the right sketch of Figure 28. In such a chain, the blocking of space by each monomer unit to be no longer occupiable by other monomer units (i.e., the self-avoidance) causes part of the volume to be **excluded volume**. As a consequence, the chain has less freedom of arrangement, and the coil must *expand*. Indeed, the self-avoiding walk in Figure 28 has a larger end-to-end distance than the self-crossing one. On top of that, if we also consider that the monomer segments of a real chain display attractive and repulsive interactions with one another and with their surrounding environment,[20] then the ratio of these monomer–monomer (M–M)

19 This expansion of our simplistic model to a more realistic one is conceptually identical to the step from the ideal gas to the real gas in elementary physical chemistry. Here, the gas molecules are also considered to have a finite volume and interactions. This is done by inclusion of two parameters for these two effects, thereby transforming the ideal gas law to the van der Waals equation.

20 In that environment, we have either solvent molecules if we consider a polymer solution or segments of other chains if we consider a polymer melt.

https://doi.org/10.1515/9783110672817-003

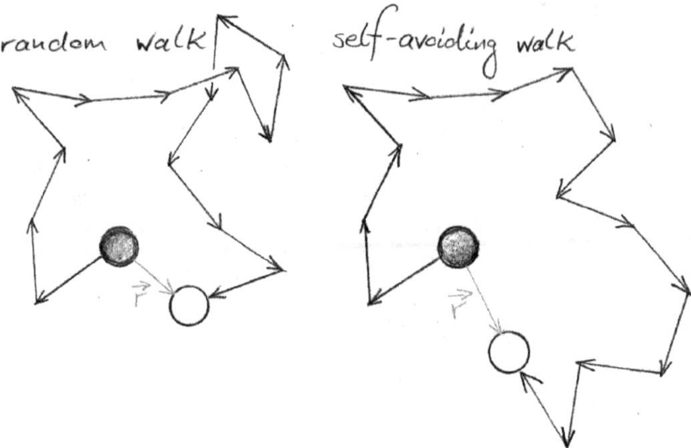

Figure 28: Trajectories of a self-crossing random walk, which corresponds to the shape of an ideal polymer coil, and a self-avoiding walk, which corresponds to the shape of a real polymer coil, in which part of the volume around each monomer unit is excluded for occupation by other units.

and monomer–solvent (M–S) attractive and repulsive interactions will further determine the extent of coil expansion; it will also determine the polymer solubility or miscibility.

3.1 Interaction potentials and excluded volume

As a start, let us examine the interaction potential between two monomer segments in a chain that have no direct chemical bond to each other (i.e., segments that are not direct neighbors along the chain). In fact, we do not even need to consider these monomers to be connected in the form of a chain; instead, it is sufficient to just consider them as molecular entities that have distance-dependent interactions through space, both attractive and repulsive. A suitable functional form to describe these interactions is the Lennard–Jones potential; it quantifies the distance-dependent interaction energy, $U(r)$, of two molecules, in a form similar to the illustration in Figure 29. This function is generally called to be a *6–12 potential*, because the energetic contribution of the attractive interactions scales with the intermolecular distance by r^{-6}, whereas the energetic contribution of the repulsive interactions scales with r^{-12}.[21] As a consequence, the repulsive interactions are more influential

21 The repulsive part of the Lennard–Jones potential reflects the strong energy penalty that arises if atoms or molecules are brought into contact so close that their occupied orbitals start to overlap; this is basically what the Pauli principle expresses, after which two electrons cannot be identical in all their quantum numbers (actually, this is expressed by a phenomenon named *exchange*

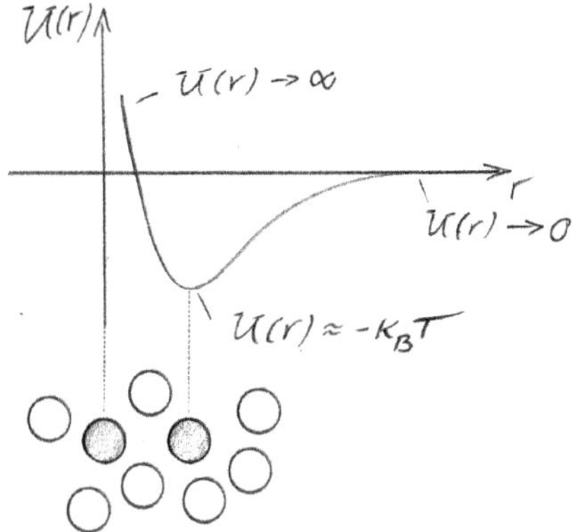

Figure 29: Effective Lennard–Jones interaction potential, $U(r)$, between two molecules with attractive effective interaction. The optimal distance, at which we have the lowest $U(r)$ value, depends on the interactions between the two molecules themselves (shaded circles) and with the surrounding medium (white circles). Picture inspired by M. Rubinstein, R. H. Colby: *Polymer Physics*, Oxford University Press, **2003**.

at short distances, where they contribute positive large $U(r)$ values. At longer distances, by contrast, the attractive interactions dominate and contribute negative $U(r)$ values (that have a greater absolute magnitude than the small positive ones contributed by the repulsive interactions there). The overlay[22] of both causes a potential well, in which a maximally negative $U(r)$ value denotes the most favorable

interaction in quantum mechanics). That strongly repulsive energy has a distance dependence that is actually not necessarily of type r^{-12}; we could also use r^{-11} or r^{-13} for it. The r^{-12} exponent is chosen somewhat arbitrarily, because it is an even number that makes a steep distance dependence. The attractive part of the Lennard–Jones potential is more substantiated. It summarizes several kinds of attractive interactions between atoms, molecules, or particles that are all based on correlated quantum fluctuations. The most significant one is spontaneous fluctuations of the dipole moment. If such fluctuations occur, even uncharged and unpolar atoms or molecules show transient dipole moments, and if two or more opposite moments form a pair or a cluster, this causes a decrease in energy (which makes it favorable) along with effective momentary attraction of the paired or clustered partners. This effect is named *London dispersion.* It is weak, but ubiquitous, and it exhibits a distance-dependent scaling of r^{-6}. (Side note: this kind of interaction is stronger the higher the polarizability of the molecules or atoms is, which, in turn, is stronger the higher the number of electrons in the shell is; this is why argon has a higher boiling point than helium.) There is two more such kinds of transient attractive interactions that also exhibit r^{-6} dependencies. All these are summarized by the term *van der Waals interactions.*

22 This is done by adding up the two parts as a sum, whereby the attractive one comes along with a negative sign, as it lowers the total energy, whereas the repulsive one comes along with a positive sign, as it increases the total energy. The exact formula is $U(r) = \varepsilon\left(\left(\frac{r_e}{r}\right)^{-12} - 2\left(\frac{r_e}{r}\right)^{-6}\right)$. In this equation, r_e is the equilibrium distance at which the potential has its minimum, and ε is the depth of the energy well at that separation. Note that the attractive term needs a numerical prefactor of 2, because only

equilibrium distance. At infinite distances, the potential levels off toward zero, because there the molecules are too far apart to "see each other".

In a real polymer system, we need to consider the attractive and repulsive interactions for two species: the monomer (M) and the surrounding medium, that is, the solvent (S). We therefore have to take into account M–M and M–S interactions, both having attractive and repulsive contributions each. To simplify the discussion, we discuss *effective* M–M interactions, in which we incorporate both the M–M and M–S contributions. We do this by redefining repulsive M–S interactions to be just the same as attractive M–M interactions, because both have the same effect: the monomers prefer to stay in closer proximity to one another than to the solvent. The same works for the opposite case: attractive M–S interactions just act like effectively repulsive M–M interactions, as in both cases, the monomers prefer to be further apart from one another than from the solvent. The resulting distance-dependent effective M–M interaction potential has a Lennard–Jones-type appearance again, as depicted in Figure 29.

Based on this premise, we can delimit three boundary cases:

1. In most cases, M–M contact is favored over M–S contact, so we have attractive effective M–M interactions and a minimum in $U(r)$, as depicted in Figure 29. This is due to the perfect structural match of two monomer units M to one another, whereas that of M and S is not as perfect, but at most just similar. As a result, whatever kind of interactions M can undergo (be it hydrogen bonding, dipolar interaction, van der Waals interaction, or whatever), it will find a perfect partner to do so in another M, and a less perfect (at most similarly good, but never as perfect) partner to do so in S, such that M–M is more preferable than M–S. In most polymers, the M–M interactions are indeed van der Waals or dipole–dipole forces, which comes along with effective M–M interaction energies in the order of $k_{\mathrm{B}}T$; this magnitude quantifies the depth of the well in the effective M–M interaction potential $U(r)$.
2. In the second boundary case, the M–M interactions are equal to the M–S interactions; this situation is encountered if M and S have practically the same chemical structure, thereby allowing them to establish the same interactions equally fine with one another or with each other. Consequently, the effective M–M interaction is zero. The $U(r)$-potential then does not exhibit any potential well, but only the hard sphere repulsive upturn in the limit $r \to 0$ (where $U(r) \to \infty$).
3. In the third boundary case, the M–M interactions are *dis*favored over the M–S interactions. This is a rare special-case scenario that we may have in polyelectrolytes in which each monomer unit carries a like charge, such that the monomer

then will the potential have a value of $-\varepsilon$ at the equilibrium distance r_{e} along with a potential minimum (i.e., a zero first derivative) at that distance.

units repel each other.[23] Again, no potential well exists in $U(r)$, but by contrast, an additional repulsive term is added on top of the inherent hard-sphere repulsion.

In the next step, we calculate the probability p to find two monomers at a certain distance r. This is simply done by deriving a Boltzmann term from the interaction energy potential $U(r)$:[24]

$$p \sim exp\left(\frac{-U(r)}{k_B T}\right) \tag{3.1}$$

A graphical representation of this term is shown in Figure 30(A) for the case of attractive effective M–M interactions. We can see that the probability to find two monomers at very short distances is zero, which corresponds to the hard-sphere

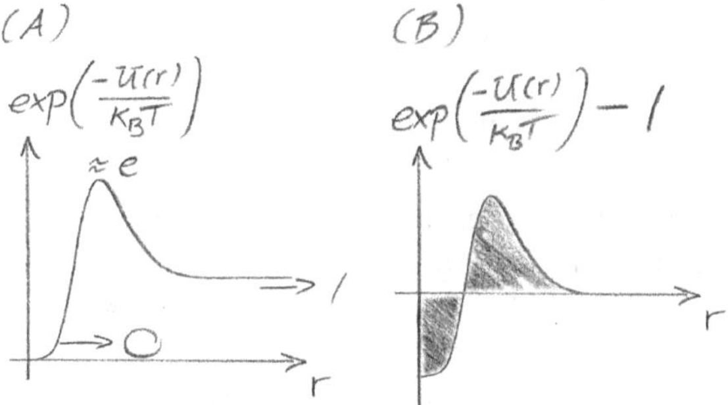

Figure 30: **(A)** Probability to find two monomers at a specific distance, r, for the case of attractive effective monomer–monomer (M–M) interactions. At short distances, the repulsive M–M interactions are so strong that the probability to find two monomers at that separation is practically zero. In contrast, the probability is highest at the $U(r)$-potential minimum, while it is independent of r for $r \to \infty$. **(B)** Plot of the *Mayer f-function*; it normalizes the Boltzmann term shown in Panel (A) to a value of zero at $r \to \infty$. From that functional form, the excluded volume can be calculated by integrating the total area underneath the graph, shaded in gray. Picture redrawn from M. Rubinstein, R. H. Colby: *Polymer Physics*, Oxford University Press, **2003**.

23 We may also have such a case if there is specific attractive M–S interaction, which manifests itself as if there were repulsive M–M interaction. This can be the case if the monomer and the solvent are somehow mutually complementary, for example, if the one carries hydrogen-bonding donor sites, whereas the other carries hydrogen-bonding acceptor sites, thereby being able to form heterocomplementary bonding with one another that does not work among themselves alone.

24 In general, the likelihood (and therewith the population or frequency of occurrence) of a state or situation of interest with a known energy U is estimated by the Boltzmann distribution: $\frac{n_U}{n} = \exp\left(\frac{-U}{k_B T}\right)$; the same principle holds for our situation of interest, which is the likelihood of having a certain distance r between two monomers if that comes along with an energy of $U(r)$.

repulsion term of the interaction energy potential, where $U(r) \to \infty$. The highest probability can be found at the most favorable distance r that corresponds to the minimum in the potential well. In this region, $U(r)$ has negative values of around $k_B T$, which translates to values of the Boltzmann term in the range of about e (=2.718). At very long distances, the probability is independent of r, because there are no operable interactions between the two monomers at these large separations. Here, $U(r)$ has a plateau value of zero, such that the Boltzmann term has a plateau value of one.

When we normalize the Boltzmann expression such that the limit for $r \to \infty$ equals zero rather than one, we generate an auxiliary function called the *Mayer f-function*, as shown in Figure 30(B):

$$f(r) = \exp\left(\frac{-U(r)}{k_B T}\right) - 1 \tag{3.2}$$

By integrating over the Mayer f-function, we can calculate the **excluded volume**, which corresponds to the area underneath the curve, shown in gray in Figure 30(B):

$$v_e = -\int f(r)\mathrm{d}r^3 = -\int 4\pi r^2 f(r)\mathrm{d}r \tag{3.3}$$

The excluded volume quantifies the space that each chain segment blocks in its surrounding due to (i) its own volume and, on top of that, (ii) the M–M interactions; if these interactions are repulsive, the blocking of (i) is further exacerbated, whereas if these interactions are attractive, the blocking of (i) is attenuated. The repulsive term in the potential $U(r)$ at distances shorter than the equilibrium distance at which the potential has its minimum, $r < r_e$, has a negative contribution to the integral over the Mayer f-function, which translates into a positive contribution to the excluded volume. By contrast, the attractive term at $r > r_e$ has a positive contribution to the integral over the Mayer f-function, and hence, a negative contribution to the excluded volume. The example in Figure 30(B) shows a situation in which the attractive and repulsive parts largely balance each other, leading to an excluded volume close to zero. This is a very special state, named the *Θ-state*, in which the chain displays a quasi-ideal conformation. That special state is very important in the field of polymer physics, and we will treat it more deeply in the following.

3.2 Classification of solvents

Based on what we have learned about M–M and M–S interactions, we can compile a list that classifies solvents by the extent of the resulting excluded volume v_e that a chain has in them.

When the M–M interactions are equal to the M–S interactions, the solvent is called to be an **athermal solvent.** This terminology is because when the interactions are the same, so is their temperature dependence, such that any change of temperature will have no effect on the effective M–M interactions. Such a solvent is structurally identical to the monomeric repeating unit of the polymer; a prime example is ethylbenzene for polystyrene, which is practically equal to polystyrene's repeating unit. In an athermal solvent, there are no attractive effective M–M interactions, leaving only the hard-sphere repulsion at short distances ($r \to 0$) in the effective M–M interaction potential $U(r)$ (where $U(r) \to \infty$). As a result, there is no positive contribution to the integral over the Mayer *f*-function, and hence, no negative contribution to the excluded volume. The outcome is a maximally positive excluded volume; it is equal to the covolume of the monomer segments: $v_e = l^3$.

In a **good solvent**, the M–M interactions are slightly more favorable than the M–S interactions. This leads to a small well in the effective M–M interaction potential $U(r)$, which somewhat compensates the inherent M–M hard-sphere repulsion. The excluded volume, thus, will still be positive, but smaller than in the athermal case: $0 < v_e < l^3$. A typical example for a good solvent is toluene for polystyrene.

A very special case is the **Θ-state**, which is present in a **Θ-solvent.** In that state, there are quite attractive effective M–M interactions,[25] leading to a pronounced well in the interaction potential $U(r)$, which just exactly balances the hard-sphere M–M repulsion at short distances in the potential, as indicated in Figure 29(B). At that constellation, the excluded volume is zero, $v_e = 0$, and the chain is in a (pseudo-)ideal state[26] and adopts the shape of a Gaussian coil with random-walk conformation. The Θ-state is therefore beloved by theorists, as it allows them to model the coil conformation by simple ideal Gaussian statistics. Experimentalists, by contrast, do not love the Θ-state, because it is highly temperature dependent and can only be present at a special temperature, the Θ-temperature.[27] This temperature marks the borderline to the nonsolvent state, such that even a slight change of temperature into the wrong direction will cause precipitation of the polymer, requiring tedious redissolution and

---

25 This means that there are quite repulsive M–S interactions, meaning that the solvent S is quite dissimilar to the monomer repeating unit M.

26 A truly ideal state is one without any interactions. A pseudo-ideal or quasi-ideal state is one with attractive and repulsive interactions at balance.

27 Again, we can find an analogy to gases: a real gas differs from an ideal one in a sense that there are attractive and repulsive interactions between the gas particles and that the gas particles have a finite covolume, which is in fact nothing else than the strongly repulsive low-distance branch of the interaction-energy potential. Both these interactions are quantified by two parameters, *a* and *b*, in the equation of state, through which the ideal gas law turns into the van der Waals equation. At a special temperature, the *Boyle temperature* T_{Boyle}, however, these interactions are at balance, such that the van der Waals equation turns back into the ideal gas law. Hence, the Boyle temperature for a gas is analog to the Θ-temperature for a polymer–solvent system.

re-equilibration before experiments can be conducted. A famous example of a Θ-state is polystyrene in cyclohexane at $T_\Theta = 34.5$ °C.

In a **bad solvent**, the M–M interactions are much more favorable than the M–S interactions, leading to strong effective M–M attraction. This causes a strong minimum in the effective M–M interaction potential $U(r)$ and a *negative* excluded volume in the range of $-l^3 < v_e < 0$. An example is ethanol for polystyrene.

In an even more extreme case, the **nonsolvent**, the M–M interactions are so much more favorable than the M–S interactions that all solvent is expelled from the polymer coil and the excluded volume becomes maximally negative: $v_e = -l^3$. An example is water for polystyrene.

The latter two cases cannot be realized in practice, because both actually lead to nondissolution and can therefore only be studied by computer simulations. They do, however, have practical applicability for preparatively working polymer chemists: turning a good or a Θ-solvent into a bad or nonsolvent, for example, by suitable change of temperature or by addition of a bad or nonsolvent excess, can serve to separate polymers from a mixture by precipitation, which is an easy means of polymer purification.

To complete the listing above, note that another special case is the one where the M–M interactions are *dis*favored over the M–S interactions. We may have that situation in polyelectrolyte solutions in which each monomer unit carries a like charge, such that the monomer units repel each other. We may also have such a case if there is specific attractive M–S interaction, for example, mutual heterocomplementary hydrogen bonding, which manifests itself as if there were repulsive M–M interactions. In that situation, just like in the athermal case, no well exists in the interaction-energy potential $U(r)$, and on top of that, there is an additional repulsive term added to the inherent hard-sphere repulsion. As a result, the excluded volume is even greater than in the athermal case: $v_e > l^3$.

3.3 Omnipresence of the Θ-state in polymer melts

The Θ-state cannot only be realized in solution at a specific temperature, but it is also always present in polymer melts, at any temperature. This can be understood by the following line of thought: consider a polymer melt in which one chain is somehow different in color, but structurally identical to the others. We can call this a "blue" chain in a matrix of "black" chains, as shown in Figure 31. Due to the chains' structural identity, this is an athermal state: M–M interactions are the same as M–S interactions, as the segments of the black "solvent" polymer chains are of the same kind as those of our blue "dissolved" chain. The blue chain, thus, has a strong tendency to expand. However, the black chains all have the same tendency. Thus, all chains in the system want to expand at the same time, and as a result, no

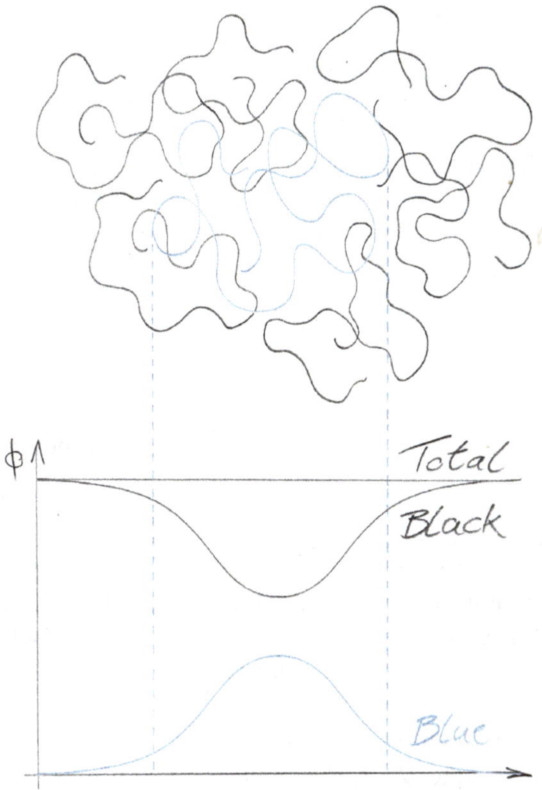

Figure 31: In a polymer melt, the excluded volume interactions in a chain of interest, here shown in blue color, are screened by overlapping segments of surrounding chains, here drawn in black color. Picture redrawn from P. G. De Gennes: *Scaling Concepts in Polymer Physics*, Cornell University Press, **1979**.

chain can actually do so. The expansion tendencies all balance each other, and the chains therefore all exhibit their unperturbed ideal conformation. In other words, the excluded volume interactions of each coil are **screened** by overlapping segments of other coils.

3.4 The conformation of real chains

LESSON 5: FLORY EXPONENT

The excluded volume introduced in the last lesson makes polymer coils expand in good solvents. In this lesson, you will learn how severe this expansion is, and how the size of a coil can be captured by a universal scaling law that introduces one of the most elementary parameters in polymer science: the Flory exponent.

3.4.1 Coil expansion

To quantify the size difference of a real polymer chain compared to an ideal one, an expansion factor α is introduced:

$$\langle R_g{}^2\rangle_{\text{real}}^{1/2} \approx \alpha \, \langle R_g{}^2\rangle_{\text{ideal}}^{1/2} \tag{3.4}$$

This factor has values above one, that is, $\alpha > 1$, for athermal and good solvents, in which the polymer coil expands. It is exactly one, which means, $\alpha = 1$, for a Θ-solvent, in which the coil has the same size as an ideal polymer chain. For bad or nonsolvents, the expansion factor is smaller than one, that is, $\alpha < 1$.

But how severe is this expansion? Let us consider polyethylene with $N_K = 50$ Kuhn-segments as an example. In Chapter 2.2.8, the characteristic ratio of this polymer was named to be $C_\infty = 6.87$ (see also Figure 17), so we get a length of $l_K = C_\infty l = 6.87 \cdot 0.154\,\text{nm} = 1.06\,\text{nm}$ for each Kuhn-segment.

At strong attractive effective M–M interactions, that is, at non-solvency conditions, the polymer is collapsed to a dense globule with volume $V = N_K l_K{}^3 = 60\,\text{nm}^3$. This corresponds to a linear dimension of $R = V^{1/3} = N_K{}^{1/3} l_K = 3.9\,\text{nm}$, as shown in Figure 32(A). Hence, in this collapsed state, the globule's across diameter is just about eight times the length of one of its Kuhn-segments (as also seen in Figure 32 (A)). This shows how densely packed the material is in this entity.

With balanced M–M interactions, or zero effective M–M interactions, we realize the Θ-state. As discussed above, the polymer then has the shape of a random Gaussian coil and obeys the ideal chain scaling law $R = N_K{}^{1/2} l_K = 7.5\,\text{nm}$, as shown in Figure 32(B). Hence, in that state, compared to the nonsolvent situation discussed earlier, the random coil is expanded to be about twice as large as the collapsed globule.

When we put the polymer chain into a good solvent, it has repulsive effective M–M interactions, which manifests itself in a positive excluded volume v_e. The coil now has the shape of a self-avoiding walk and obeys scaling according to $R = N_K{}^{3/5} l_K = 11\,\text{nm}$, as shown in Figure 32(C). (This scaling law will be derived in the next section.) Now, the polymer has a size right in the typical colloidal domain.

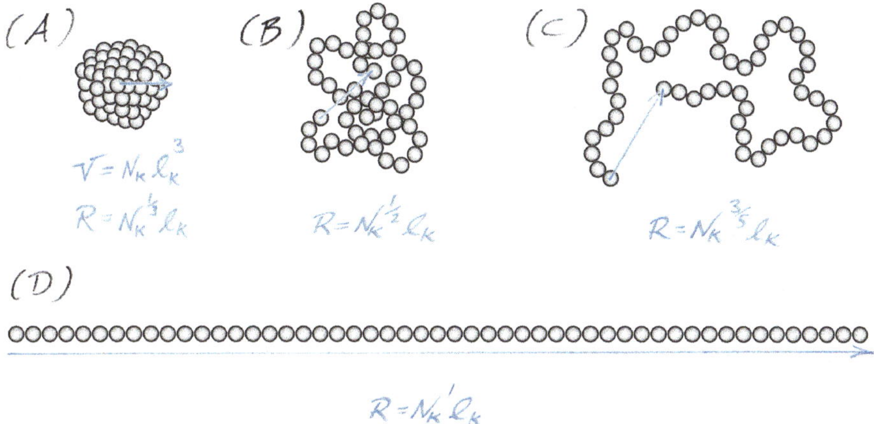

Figure 32: Visualization of the different stages of expansion for a polymer with 50 Kuhn-segments. **(A)** In its fully collapsed state, the polymer is a dense globule ($R = 3.9$ nm), whereas **(B)** at Θ-conditions, it is a coil with the shape of a random walk ($R = 7.5$ nm). **(C)** Effective repulsive M–M interactions cause the coil to expand to a self-avoiding walk shape ($R = 11$ nm). **(D)** At maximally strong repulsive M–M interactions, the polymer fully expands to a stiff rod ($R = 53$ nm). Volume and size dependencies on the number of Kuhn segments, N_K, and their length l_K, for each regime are shown underneath each image.

At maximally strong repulsive effective M–M interactions, meaning at a maximally large excluded volume v_e, the chain is in its most expanded conformation, which is a rodlike object, as shown in Figure 32(D). Its size can be calculated according to $R = N_K l_K = 53$ nm, which is considerably larger than in the other states discussed before.

We see from the above example that a better solvent quality leads to remarkable expansion of the polymer coil, from a few single to several tens of nanometers. This, however, comes at the cost of a lower segmental density in the coil interior. Throughout our discussion above, we have treated the same polymer chain; we have not added new segments to it, but only given it more space to arrange itself. When we do this, we get a greater coil expansion, but, in turn, we naturally reduce the number of monomer segments per volume. This is shown for the segmental density of polystyrene in either the θ-state or in a good-solvent state in Figure 33.

Generally, we can appraise the polymer coil's size for all expansion regimes by its root-mean-square end-to-end distance according to the following general scaling law:

$$\langle r^2 \rangle^{1/2} \approx N^{\nu} l \tag{3.5}$$

In this law, ν is the **Flory exponent** that varies between 1/3 and 1 in our above line of thought, depending on the solvent quality. It is this power-law exponent that causes the difference in the size of a polymer at different states of solvency, because

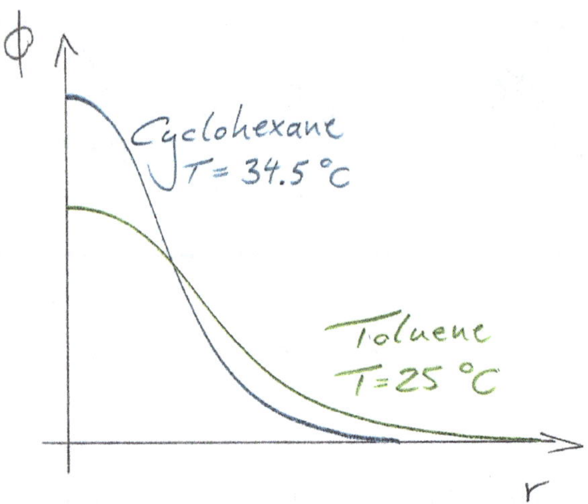

Figure 33: Segmental density of a polystyrene coil in a good and a Θ-solvent. The good solvent expands the coil as compared to the Θ-solvent, leading to a greater occupancy of the volume far from the coil center, but in turn, to a lower segmental density in its core.

in the power law (3.5), a large ν empowers a given number of (Kuhn-)segments N to a greater extent than a small ν does. Due to the mathematical nature of power laws, this different empowerment is even more pronounced when the number of segments in the chain is large. For example, if we consider polyethylene again, but this time with 20,000 repeating units (just as we have done in Section 2.2.8), we have a polymer with $N_K = \frac{N}{C_\infty} = 2899$ Kuhn-segments of length $l_K = C_\infty l = 1.06\,\text{nm}$ each. In a fully collapsed state, this polymer has a size of $R = N_K{}^{1/3} l_K = 15\,\text{nm}$, whereas at Θ-conditions, it has a size of $R = N_K{}^{1/2} l_K = 57\,\text{nm}$ (just as already calculated in Section 2.2.8). In a good solvent, it is expanded even further, to a size of $R = N_K{}^{3/5} l_K = 127$ nm, and when fully expanded to a rod, the polymer exhibits a size of $R = N_K{}^{1} l_K = 3080$ nm (as also calculated already in Section 2.2.8). The size of this polymer differs even more markedly than the one in the example of Figure 32 depending on its solvency state, spanning the whole colloidal domain between single nanometers to a few micrometers.

3.4.2 Flory theory of a polymer in a good solvent

Let us consider an expanded polymer coil composed of N monomer units and a size R bigger than that of the ideal Gaussian coil, $R > R_0 = N^{1/2} l$.

We can estimate the number of monomers that are located inside the excluded volume of a given first one as follows:

$$v_e \cdot \frac{N}{R^3} \tag{3.6}$$

Here, v_e is the excluded volume of one monomer segment, which is the volume we want to consider here, and $\frac{N}{R^3}$ is the number of segments in the coil per volume of the coil, in other words, the general segmental density in the coil. In the optimal case, the expression of eq. (3.6) should equal one. Then, the excluded volume of a given monomer segment is only occupied by one such segment: by itself. Any additional monomers that intrude into the excluded volume impose an energy penalty that can be appraised to be k_BT per such unfavorable M–M contact:

$$F_{\text{excl, per monomer}} = k_B T v_e \frac{N}{R^3} \tag{3.7}$$

If there is only one monomer in the excluded volume (which is the optimal situation), the resulting energy is k_BT, which can be easily "paid" by the ever-present thermal energy that is exactly k_BT. Each additional monomer–monomer contact in the excluded volume, however, adds another k_BT-increment to the penalty. Consequently, additional energy is required. This is generally unfavorable, such that the system has a tendency to avoid such M–M contacts, which can be achieved by coil expansion. Upon such an expansion, R in eq. (3.7) will get larger, such that $F_{\text{excl,per monomer}}$ will get lower, which means it is less unfavorable.

In a chain with N monomers, the excluded-volume-interaction energy appraised by eq. (3.7) is present N times:

$$F_{\text{excl, per chain}} = k_B T v_e \frac{N^2}{R^3} \tag{3.8}$$

Again, the only nonconstant parameter here that can be adjusted is the polymer coil's size R. Because R enters the equation in the denominator, the chain can minimize this energy contribution by *increasing* its size, $R \to \infty$. The coil, thus, has an *expansion tendency.*

On the other hand, coil expansion causes an entropy-based restoring force, corresponding to an energy F_{elast}, due to the loss of microconformational freedom upon coil expansion, as we have learned in Section 2.6. This can be appraised as

$$F_{\text{elast}} = F_0 + \frac{3}{2} \cdot \frac{k_B T R^2}{N l^2} \tag{3.9}$$

Again, the size R is the only adjustable parameter that, in this case, enters the equation in the numerator. From this, it follows that the chain can minimize this energy contribution by *decreasing* its size, $R \to 0$. The coil, thus, has a *contraction tendency.*

To calculate the total energy of the chain subject to the two preceding influences, we have to sum up both the excluded-volume-interaction energy F_{excl} and the elastic energy F_{elast}:

$$F = F_{\text{excl}} + F_{\text{elast}} = k_B T\left(v_e \frac{N^2}{R^3} + \frac{3R^2}{2Nl^2}\right) \tag{3.10}$$

To find the coil size with the minimum total energy, we calculate the derivative and set it zero:

$$\frac{\partial F}{\partial R} = k_B T\left(-3v_e \frac{N^2}{R^4} + \frac{3R}{Nl^2}\right) \stackrel{!}{=} 0 \Rightarrow R \sim {v_e}^{1/5}\, l^{2/5}\, N^{3/5} \tag{3.11}$$

With that, we have shown that the *N*-dependent scaling of *R* of a coil in a good solvent is

$$R \sim N^{3/5} \tag{3.12}$$

There is just one problem: so far, we always found a length-dependent scaling of $R \sim l^1$. This is in contradiction to eq. (3.11), where we have found $R \sim l^{2/5}$ instead. The reason for this discrepancy is a mistake in the above derivation. In eq. (3.9), we have used ideal-chain scaling of $\langle r^2 \rangle = Nl^2$ in the denominator, even though all our above line of argument in fact has the purpose to appraise *nonideal*, expanded coil dimensions. This mistake causes the erroneous finding of $R \sim l^{2/5}$. Nevertheless, our finding of $R \sim N^{3/5}$ is correct though. Why is that? It is because in view of the *N*-dependence, our mistake is compensated by another one. In eq. (3.6), we have appraised the segmental density in the coil to be uniform (in the form of N/R^3), whereas we know from Section 2.3 that it actually has a Gaussian radial profile. This wrong estimate of the segmental density in eq. (3.6) compensates the incorrect scaling in the denominator of eq. (3.9) in view of the *N*-dependence of *R*, whereas it is not cancelled out in view of the *l*-dependence of *R*.

The general scaling of *R* as a function of *N* in the form of eq. (3.12) can be written as

$$R = \langle \vec{r}^2 \rangle^{1/2} \sim N^{\nu} \tag{3.13}$$

with ν the Flory exponent. For a polymer in a good solvent, we have just derived it to be $\nu = 3/5$. For an ideal chain, that is, a chain at θ-conditions, we have shown earlier (in Section 2.2, eq. (2.5)) that $\nu = 1/2$. For a coil that is fully collapsed to a dense globule, that is, a chain in a nonsolvent, we have shown earlier (Section 3.4.1) that $\nu = 1/3$, whereas for the other extreme, a fully expanded rodlike chain, we have shown that $\nu = 1$.

If we conduct the preceding estimate for the general case of a *d*-dimensional space, we have to use R^d in eqs. (3.6)–(3.8) and a numerical factor of $d/2$ in eq. (3.9). With that, we get

$$R = \langle \vec{r}^2 \rangle^{1/2} \sim N^{3/(d_{\text{geometrical}}+2)} \tag{3.14}$$

In this general form, the Flory exponent is $\nu = 3/(d_{\text{geometrical}} + 2)$, allowing us to illustrate the role of the geometrical dimension, $d_{\text{geometrical}}$, as follows. In earlier sections, we have considered the three-dimensional case, $d = 3$, where the Flory exponent is 3/5. In a two-dimensional situation, $d = 2$, this is little different. Here, according to eq. (3.14), the Flory exponent has a value of 3/4, which is larger than 3/5, indicating that a coil with given N has a greater R in two dimension than in three dimension. The reason is that there is less freedom for the polymer coil to arrange itself in a 2*d*-space than in a 3*d*-space, because it has only two spatial directions to occupy. Hence, the coil must expand more in 2*d* than in 3*d* to avoid unfavorable M–M contacts. This trend is even more extreme in a one-dimensional situation, $d = 1$. Now, according to eq. (3.14), the Flory exponent is 3/3 = 1. This is because in 1*d*, the coil has no other way to avoid M–M contacts than to fully expanding itself into a rodlike object with $R = Nl$. By extreme contrast, in a four-dimensional situation, $d = 4$, the coil has so much freedom to arrange itself that M–M contacts are generally very unlikely. The coil can therefore adopt its random Gaussian shape like an ideal chain or a real chain in the Θ-state. Here, consequently, the Flory exponent is 3/6 = 1/2. (If we further follow this line of thought, even higher geometrical dimensions such as $d = 5, 6, \ldots$ would denote smaller Flory exponents such as ν= 3/7, 3/8, . . ., indicating that a coil in such high dimensions would contract. This can be understood on the very same basis as the discussion just led: the higher the geometrical dimensions, the smaller can be the coil size without causing unfavorable M–M contacts. Higher dimensions, therefore, allow R to be small, which minimizes the elastic energy according to eq. (3.9) (where R enters in the form of R^2, irrespective of d), without excessively increasing the excluded volume interaction energy according to eq. (3.8) (where R enters in the form of R^{-d}, which means that if d is high, there is lesser need for large R to make that term small).)

In addition to the preceding discussion of the geometrical dimension, we may also do so for the fractal dimension, a concept that we have introduced in Section 2.6.2. With this dimension, we get

$$R = \langle \vec{r}^2 \rangle^{1/2} \sim N^{1/d_{\text{fractal}}} \tag{3.15}$$

Hence, the Flory exponent is nothing else than the inverse fractal dimension. When looking on the compilation of Flory exponents in one to four geometrical dimensions above, we see that the inverse of these exponents, that is, the fractal dimension, is getting lower and lower. In general, a smaller fractal dimension denotes the object that it belongs to be less dense. If we compare the fractal dimension in four geometrical dimensions, $d_{\text{fractal}} = 1/\nu = 1/(1/2) = 2$, to that in three geometrical dimensions, $d_{\text{fractal}} = 1/\nu = 1/(3/5) = 5/3$, and in two geometrical dimensions, $d_{\text{fractal}} = 1/\nu = 1/(3/4) = 4/3$, we obtain a series of smaller and smaller values. This means that chains in lower geometrical dimensions are less dense than in higher geometrical dimensions. The reason for that is the excluded-volume repulsion between the monomer segments, which pushes them apart from one another to avoid M–M contact; this does not need to be as pronounced in higher geometrical dimensions, as such contact is generally less likely there. A special case is the one of a densely collapsed globular polymer in a nonsolvent; here we have a Flory exponent of $\nu = 1/3$, corresponding to a fractal dimension of $d_{\text{fractal}} = 1/\nu = 1/(1/3) = 3$. This fractal dimension matches the geometrical one, thereby denoting a nonfuzzy, dense object, which a collapsed polymer globule indeed is.

Paul John Flory was born on June 19, 1910, in Sterling, Illinois. He studied at Manchester College and obtained his Bachelor of Science degree during the time of the great depression, supporting himself with various jobs at the side. During that period, his interest in science, particularly chemistry, was inspired by Professor Carl W. Holl, who encouraged him to enter graduate school at Ohio State University in 1931. Flory followed this advise, pursued his graduate studies at Ohio State, and obtained his PhD degree in 1934. From 1933 to 1948, he worked in several industrial research laboratories for companies such as DuPont, Standard Oil, and Goodyear. He was offered a faculty position at Cornell University in 1948, where he stayed until 1957. Having transferred to and led the Mellon Institute in Pittsburgh until 1961, he became a full professor at Stanford University until he retired in 1974. One year prior to his retirement, he received the Nobel Prize in Chemistry "for his fundamental achievements, both theoretical and experimental, in the physical chemistry of macromolecules". He died on September 9, 1985, at the age of 75 in Big Sur, California.

Figure 34: Portrait of Paul J. Flory. Image reproduced with permission from Stanford University Libraries, Department of Special Collections and University Archives (SC0122, Stanford University News Service records, Box 90, Folder 48, Paul Flory.).

3.5 Deformation of real chains

Just like in Chapter 2, after having made ourselves a mind about the size and shape of a real chain, we now want to translate that structural information into information on properties, specifically, into the elastic properties upon deformation of the chain. Other than for the simple case of an ideal chain, as treated in Section 2.6, such a discussion for a real chain will be mathematically challenging. To simplify it, Rubinstein's and Colby's clever **scaling argument** that has been used already in

Section 2.6.1 can be applied again to conceptually renormalize the polymer chain within a *blob concept*. This approach is valid for both ideal (see Section 2.6.1) and real chains, so both are treated simultaneously below.

The root-mean-square end-to-end distance of a polymer chain is calculated as

$$\text{Ideal chain:}\ \langle r^2\rangle_0^{\,1/2} = R_0 = N^{1/2} l \tag{3.16a}$$

$$\text{Real chain:}\ \langle r^2\rangle^{1/2} = R_F = N^{3/5}\,1 \tag{3.16b}$$

Due to the self-similarity of polymer chains, the same scaling also applies to subsections of the chain that only encompass n monomers:

$$\text{Ideal chain:}\ r_0 = n^{1/2} l \tag{3.17a}$$

$$\text{Real chain:}\ r_F = n^{3/5} l \tag{3.17b}$$

We now regard a special subsection scale, named ξ. On scales smaller than ξ, the external deformation energy is weaker that the ever-present thermal noise $k_B T$; as a result, on scales smaller than ξ, the chain segments show unperturbed random-walk-type (in the case of an ideal chain) or excluded-volume-expanded (in the case of a real chain) conformations, but they do not "feel" any external deformation. By contrast, on scales larger than ξ, external deformation is effective, as its energy is stronger than $k_B T$ there. Hence, although the chain segments on scales smaller than ξ are not affected by external deformation, the whole chain is. It can therefore be conceived as an oriented sequence of *deformation blobs* of size ξ, inside of each being a chain subsegment of g monomers that is not affected by the deformation, as shown in Figure 35. In this way, the chain can maximize its entropy even under the external constraint of deformation. By adaption of eq. (3.17) to the blob scale, the blob size is expressed as

$$\text{Ideal chain:}\ \xi = g^{1/2} l \tag{3.18a}$$

$$\text{Real chain:}\ \xi = g^{3/5} l \tag{3.18b}$$

On length scales larger than ξ, the chain is an oriented sequence of deformation blobs; as a result, its end-to-end distance can be approximated as the blob size, ξ, times the number of blobs, (N/g), which we may then rewrite further by using eq. (3.18) (rearranged to g and then replacing g in N/g) and eq. (3.16) (replacing the then occurring numerators of type $Nl^{1/\nu}$)

$$\text{Ideal chain:}\ R_f \approx \xi\frac{N}{g} = \frac{Nl^2}{\xi} = \frac{R_0^{\,2}}{\xi} \tag{3.19a}$$

$$\text{Real chain:}\ R_f \approx \xi\frac{N}{g} = \frac{Nl^{5/3}}{\xi^{2/3}} = \frac{R_F^{\,5/3}}{\xi^{2/3}} \tag{3.19b}$$

Rearranging these equations yields an expression for the blob size ξ

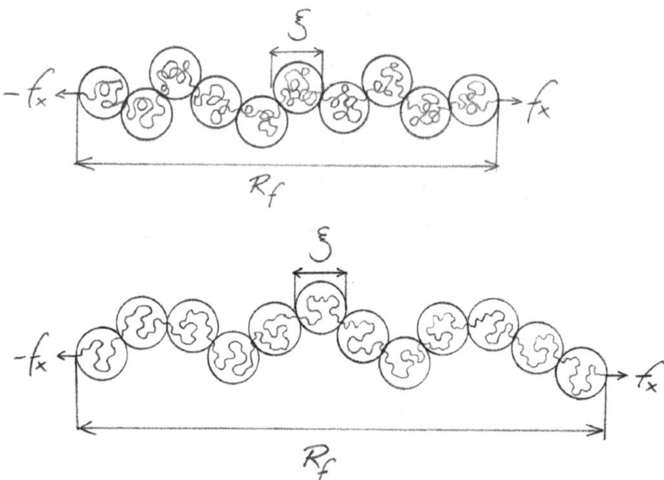

Figure 35: Modeling of a polymer chain, ideal (upper sketch) and real (lower sketch), subject to an external stretching force that leads to an end-to-end distance R_f as a conceptual object composed of blob elements with size ξ. At scales below ξ, the polymer chain does not experience deformation and displays ideal random-walk-type (ideal chain) or expanded (real chain) conformations. At scales above ξ, by contrast, in both cases, the polymer is an oriented sequence of blobs, as on these scales, the external deformation is effective. Picture redrawn from M. Rubinstein, R. H. Colby: *Polymer Physics*, Oxford University Press, **2003**.

$$\text{Ideal chain: } \xi = \frac{R_0{}^2}{R_f} \tag{3.20a}$$

$$\text{Real chain: } \xi = \frac{R_F{}^{5/2}}{R_f{}^{3/2}} \tag{3.20b}$$

As we have learned in Section 2.6.1, the free energy of deformation, F, is k_BT per blob. By using eqs. (3.19) and (3.20) to further rewrite, we get

$$\text{Ideal chain: } F_{\text{ideal}} = k_BT\frac{N}{g} = k_BT\frac{R_f}{\xi} = k_BT\left(\frac{R_f}{R_0}\right)^2 \tag{3.21a}$$

$$\text{Real chain: } F_{real} = k_BT\frac{N}{g} = k_BT\frac{R_f}{\xi} = k_BT\left(\frac{R_f}{R_F}\right)^{5/2} \tag{3.21b}$$

We have also learned that the force needed to deform the chain by a distance of R_f corresponds to the ratio of the thermal energy k_BT and the blob size ξ, which we can be expressed by eq. (3.20) to get

$$\text{Ideal chain: } f = \frac{k_BT}{\xi} = \frac{k_BT}{R_0{}^2}R_f = \frac{k_BT}{R_0} \cdot \frac{R_f}{R_0} \tag{3.22a}$$

$$\text{Real chain: } f = \frac{k_B T}{\xi} = \frac{k_B T}{R_F{}^{5/2}} R_f{}^{3/2} = \frac{k_B T}{R_F} \cdot \left(\frac{R_f}{R_F}\right)^{3/2} \tag{3.22b}$$

From this expression we realize that the force needed to deform a real chain increases stronger with R_f than the force needed to deform an ideal chain. The absolute force values, however, are always smaller for a real chain due to their bigger swollen, or "prestretched", original shape. This fact is visualized in Figure 36. A general equation for the deformation of a polymer chain can be expressed using the Flory exponent ν:

$$F = k_B T \left(\frac{R_f}{N^{\nu} l}\right)^{\frac{1}{1-\nu}} \tag{3.23}$$

To summarize, the preceding scaling discussion has shown that both ideal and real chains lose conformational freedom upon deformation, which is reflected by their entropic spring constant $k_B T/\xi$. However, they do so in different ways: an ideal chain is deformed from its Gaussian coil dimensions, R_0, whereas the real chain is already prestretched due to excluded volume interactions to a bigger size, R_F. From this, it follows that the deformational force for real chains is smaller than that of their ideal counterparts, albeit it increases more steeply upon deformation.

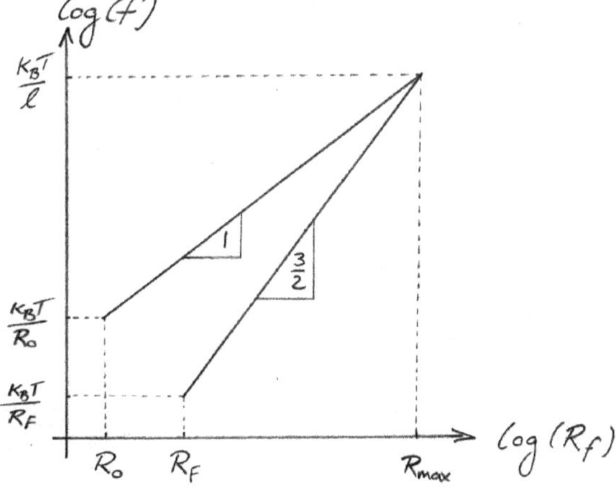

Figure 36: Force, f, needed to deform an ideal and a real polymer chain by a distance R_f. For an ideal chain, the required force scales with the deformation distance R_f with a power law exponent of one, as expressed by eq. (3.22a), whereas for a real chain, the required force scales with the deformation distance R_f with a power law exponent of 3/2, as expressed by eq. (3.22b). Thus, the force needed to deform a real chain increases more steeply than its counterpart for an ideal chain. The absolute force values, however, are always smaller for a real chain due to their bigger swollen, or "prestretched", original shape. Picture redrawn from M. Rubinstein, R. H. Colby: *Polymer Physics*, Oxford University Press, **2003**.

3.6 Chain dynamics

LESSON 6: CHAIN DYNAMICS

So far, polymer chains were viewed to be static in this book. This chapter will go beyond that view and introduce two fundamental frameworks to model and quantify chain dynamics: the Rouse and the Zimm model. Both will show that it takes a certain time before a chain can move as a whole, whereas below, only parts of it move. This time delimits the scale on which a polymer is a viscous fluid from the scale where it is a viscoelastic body.

3.6.1 Brownian motion and diffusion

In the last chapters, we have looked at the shape of ideal and real polymer chains; we have also discussed how their shape changes when they are deformed. In this chapter, we will focus on the motion, the *dynamics*, of polymer chains. As a fundament for that, we first recapitulate some elementary physical chemistry. The basic description of the thermal motion of any (molecular or colloidal) object is that of a random walk, a concept that we have already discussed in Section 2.4.1. The trajectory of such a random walk is displayed in Figure 37.

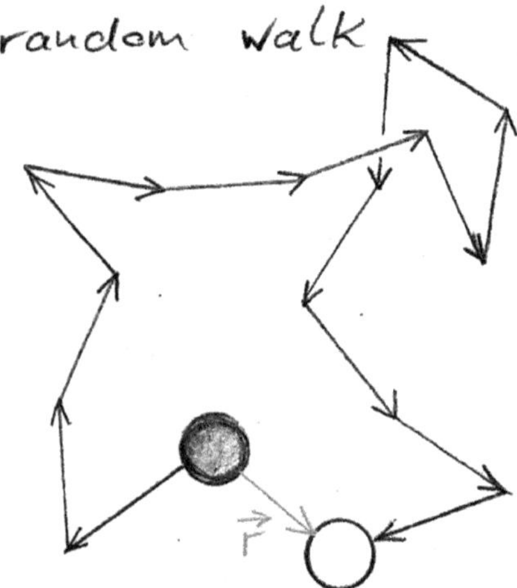

Figure 37: Trajectory of a (two-dimensional) random walk.

A characteristic quantity for random walks is their **mean-square displacement**, $<\vec{R}^2>$, which is expressed by the **Einstein–Smoluchowki equation:**[28]

$$\langle \vec{R}^2 \rangle = \langle (\vec{r}(t) - \vec{r}(0))^2 \rangle = 2\ d\ D\ t \tag{3.24}$$

Here, d denotes the geometrical dimension, and D is the translational diffusion coefficient, a quantity that expresses how mobile the moving molecule or particle is. According to eq. (3.24), D has the unit $m^2 \cdot s^{-1}$, thereby quantifying what mean-square distance the moving object passes per time. The diffusion coefficient can be calculated by the **Einstein equation**:

$$D = \frac{k_B T}{f} \tag{3.25}$$

This equation relates D to the ratio of the thermal energy that drives the diffusion, $k_B T$, and the friction that drags the diffusion; the latter is expressed by a friction coefficient, f, that connects the frictional force to the velocity of the moving object, $\vec{f} = f\vec{v}$. The friction coefficient of a spherical object is given by **Stokes' law** as

$$f = 6\pi\eta r_h \tag{3.26}$$

Here, η expresses the viscosity of the surrounding medium (this is what actually exerts the friction) and r_h denotes the hydrodynamic radius, which is the radius of the moving object itself plus its solvent shell (and potential swelling medium inside) that is dragged with it during the motion.

Both the latter equations can be combined to give the **Stokes–Einstein equation**:

$$D = \frac{k_B T}{6\pi\eta r_h} \tag{3.27}$$

Equation (3.27) is fundamental in physical chemistry, as it relates the size of a moving molecule or particle, denoted by r_h, to its mobility, denoted by D. Together with the Einstein–Smoluchowki equation (3.24), this allows us to determine how far a diffusing molecule or particle of given size r_h can move in a prescribed time t, or

28 In Figure 37, the displacement of the moving particle corresponds to the distance of the last arrow of the sequence to the starting point of the walk. The *mean-square* of it is obtained by averaging over many of such walks, whereby in that averaging, each displacement is first squared to get rid of the directional dependence. Otherwise, if that was not done, the average would always be zero, as each displacement would be cancelled by one showing exactly into the opposite direction in a great ensemble of walks. As a result, the quantity of interest is the *mean-square displacement* – the mean over many individual displacements in a squared form. Often, to relinearize the physical dimension, people further take the square-root, thereby obtaining the *root-mean-square displacement*. We also do so all over Chapters 2 and 3 when we talk about the root-mean-square end-to-end-distance of polymer chains.

vice versa, how long it takes to move a given distance R^2. This is of elementary relevance in many scientific fields, for example, in the field of drug delivery, when the time shall be appraised that a drug will need to reach a receptor in a cell or tissue environment, or vice versa, when it shall be appraised how far at all the drug can move in a given timeframe. The same question is also relevant in fields like chemical technology when it comes to appraising the efficiency of heterogeneous reactions, where partners must find each other by diffusion across phase boundaries, or in chemical engineering, when it comes to appraising how far diffusive smearing will impair precision in micropatterning or 3*d* printing.

In the field of polymer and colloid science, a specified form of eq. (3.24) is relevant; this is a form in which eq. (3.24) is rephrased such to express the timescale τ for the displacement of a moving (macro)molecule or (colloid)particle by exactly its own size, R:

$$\tau = \frac{R^2}{2d \cdot D} = \frac{R^2 f}{2dk_B T} \tag{3.28}$$

On timescales shorter than this characteristic time τ, the colloidal or polymeric building blocks of a material cannot move over distances at least corresponding to their own size, meaning they are practically static. By contrast, on timescales longer than τ, the material's building blocks are macroscopically mobile. As a result, τ delimits the time domain on which a material is a solid from that on which it is a liquid. For polymeric and colloidal matter, this limiting time is often on experimentally relevant orders of magnitude, which causes these materials to exhibit both solidlike and liquidlike appearance, depending on the timescale of observation.

So far, all the above discussion accounts for simple molecules or particles. When it comes to the dynamics of a flexible polymer coil, however, in addition to its global motion, we also must consider that it has multiple kinds of coil-internal dynamics, as it is a large multibody object. To account for this complexity, two different models have been developed: the **Rouse model** and the **Zimm model.**

3.6.2 The Rouse model

The Rouse model conceptually describes a polymer chain as a number of N spherical beads, representing the monomeric units, connected through elastic springs of length l, representing the bonds between the monomer segments, as depicted in Figure 38. Each bead is assigned an individual segmental friction coefficient $f_{segment}$. Together, the beads constitute a **freely drained coil**, which means that only the beads feel friction with the surrounding medium, but the springs do not. As a result, the solvent can pass freely through the polymer coil and hit each bead, where it imparts a frictional increment $f_{segment}$. The total friction coefficient of the coil is therefore simply the sum of all these individual frictions: $f_{total} = N \cdot f_{segment}$. This generates a system of

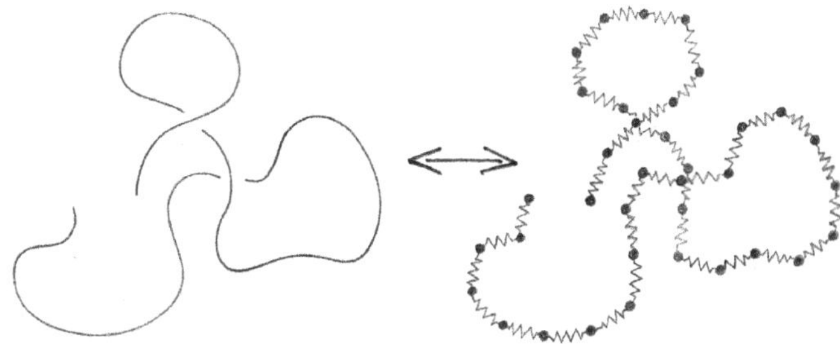

Figure 38: Spring-and-bead modeling of a chain in the Rouse model. Picture redrawn from M. Rubinstein, R. H. Colby: *Polymer Physics*, Oxford University Press, **2003**.

coupled differential equations from the equations of motion of the beads subject to drag by the springs and friction by the solvent. If we substitute f_{total} in the Einstein eq. (3.25) by $N \cdot f_{\text{segment}}$, we get

$$D_{\text{Rouse}} = \frac{k_B T}{f_{\text{total}}} = \frac{k_B T}{N \cdot f_{\text{segment}}} \sim N^{-1} \tag{3.29}$$

We now formulate an Einstein–Smoluchowski analog expression for the timescale to achieve a displacement by exactly one coil size, the **Rouse time**:

$$\tau_{\text{Rouse}} = \frac{R^2}{2d \cdot D_{\text{Rouse}}} = \frac{f_{\text{segment}}}{2dk_B T} NR^2 \tag{3.30}$$

This time denotes the upper extreme of a whole spectrum of relaxation times that we will discuss in more detail below. On timescales longer than τ_{Rouse}, the coil can migrate over distances further than its own size; macroscopically, this means that on these timescales the building blocks of a polymer material can effectively displace against each other, such that the material exhibits flow.

The other extreme of the characteristic relaxation time spectrum is the time it takes for at least each monomeric segment to be displaced by a distance equal to its own size, l:

$$\tau_0 = \frac{f_{\text{segment}} \cdot l^2}{2dk_B T} \tag{3.31}$$

On timescales shorter than τ_0, no motion whatsoever is possible in the coil. On these short timescales, the material therefore is an (energy-)elastic solid with glassy appearance.

A polymer chain is a fractal object with size $R = N^{\nu} l$ (cf. Section 2.6.2). When we insert this into the above equation for the Rouse time, we can derive a generalized expression that relates τ_{Rouse} to τ_0:

$$\tau_{\text{Rouse}} = \frac{f_{\text{segment}} \cdot l^2}{2dk_{\text{B}}T} N^{1+2\nu} = \tau_0 N^{1+2\nu} \tag{3.32}$$

The latter equation combines the two characteristic ends of the **relaxation time spectrum**, and thereby delimits three different domains of dynamics. On timescales shorter than τ_0, that is, $t < \tau_0$, the polymer does not have sufficient time to move at all, neither the entire chain nor any of its constituent segments. On this timescale, the polymer forms an energy-elastic, *glassy solid*. On intermediate timescales, $\tau_0 < t < \tau_{\text{Rouse}}$, at least monomeric segments and sequences of them (i.e., subchain segments) can move over distances equal to their own size. The time is, however, not yet long enough for the entire chain to effectively displace itself. On this timescale, the polymer is a *viscoelastic solid*. Only on timescales longer than the Rouse time, $t > \tau_{\text{Rouse}}$, the whole polymer coil can move over distances greater than its own size. Now, the material exhibits flow and is a *viscoelastic liquid*.

3.6.3 The Zimm model

So far, we have considered the polymer coil to be freely drained by solvent. This view, however, is not always true. M–S interactions can cause moving polymer chain segments to drag solvent molecules with them. These, in turn, do the same to the solvent molecules adjacent to them. This drag spreads from one solvent molecule to another until, eventually, it reaches other segments of the polymer chain. Hence, even distant segments are coupled to each other through space by **hydrodynamic interactions.** As a result, each coil drags the solvent in its pervaded volume with it on its motion; this trapped solvent is in diffusive exchange with the medium in the surrounding, but does not drain through the coil. As a result, the coils appear like solvent-filled nanogel particles, as visualized in Figure 39.

The *Zimm model* expands on the Rouse model to incorporate such hydrodynamic interactions. It assumes that the polymer coil and the trapped solvent inside act together as a united object of size $R = N^\nu l$. Substitution of the hydrodynamic radius, r_{h}, in the Stokes–Einstein equation with this expression yields

$$D_{\text{Zimm}} = \frac{k_{\text{B}}T}{6\pi\eta r_{\text{h}}} \approx \frac{k_{\text{B}}T}{\eta N^\nu l} \sim N^{-\nu} \tag{3.33}$$

Compared to the Rouse model, the power-law scaling is less steep here: it is just $-\nu$ compared to -1.[29] Hence, in the Zimm model, the diffusion coefficient has a less

29 Note: for stiff, rodlike chains, with $\nu = 1$, the Zimm-type scaling matches the Rouse-type scaling, as for such a polymer, there is no trapped solvent in it, as it is fully decoiled. So, Zimm- and Rouse-type dynamics become indistinguishable for this kind of chain.

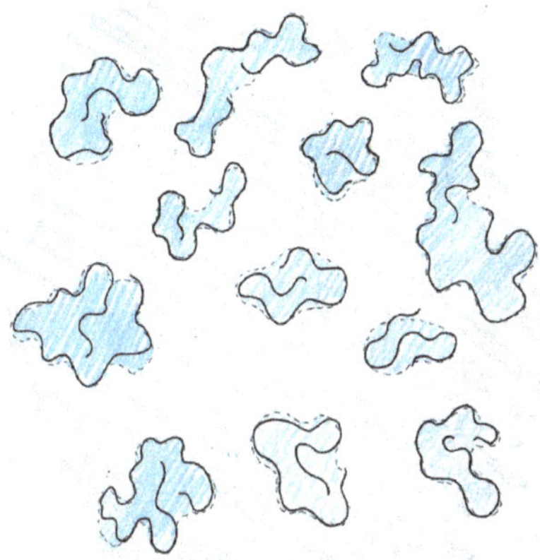

Figure 39: Modeling of polymer coils as microgels entrapping the portion of solvent within their pervaded volume due to hydrodynamic interactions and dragging it with them on their motion through the free solvent in the surrounding medium. The trapped portion of solvent is in diffusive exchange with that in the surrounding, but the coils are not drained by the medium. Picture inspired by B. Vollmert: *Grundriss der Makromolekularen Chemie*, Springer, **1962**.

pronounced dependence on N than in the Rouse model. The reason is the lack of solvent draining in the Zimm model. In the Rouse model, viscous drag is exerted by each bead that constitutes the coil; in the Zimm model this is done only by the beads (and the trapped solvent between them) on the coil's frontal face. Thus, any change of the number of beads directly translates to the drag on the coil motion in the Rouse model, and hence, results in an inverse proportionality of its diffusion coefficient on N, whereas this is less pronounced in the Zimm model.

As we have done in the Rouse model, we can formulate an Einstein–Smoluchowski analog expression for the timescale of displacement by exactly one coil size, the **Zimm time**:

$$\tau_{\text{Zimm}} = \frac{R^2}{2d \cdot D_{\text{Zimm}}} \approx \frac{\eta}{k_B T} R^3 \approx \frac{\eta l^3}{k_B T} N^{3\nu} \approx \tau_0 N^{3\nu} \tag{3.34}$$

Again, compared to the Rouse time, the power law exponent is smaller here: 3ν compared to $(1 + 2\nu)$.[30] Again, this means that there is a weaker dependence of the longest relaxation time τ on the number of monomer segments N in the Zimm model than in the Rouse model. As a consequence, τ_{Zimm} is shorter than τ_{Rouse}. The reason is the absence of solvent draining, which comes along with less viscous drag, causing the time its takes for the coil to diffuse a given distance (such as that

30 Yet again, for stiff, rodlike chains, with $\nu = 1$, the Zimm-type scaling matches the Rouse-type scaling.

of its own size) to be shorter than in the case of more pronounced viscous drag in the Rouse scenario.

Bruno Hasbrouck Zimm was born on October 31, 1920, in Woodstock, New York. He studied at Columbia University, where he obtained his bachelor's degree in 1941, his master's degree in 1943, and his PhD degree under Joseph. E. Mayer in 1944. He then moved across town for postdoctoral work with Herman Mark at the Polytechnic Institute of Brooklyn. In 1946, he transferred to the University of California in Berkeley, where he became an assistant professor in 1950–1952. After that, he was the head of the General Electric research labs in Schenectady. In 1960, he became a full professor at the University of California in San Diego. He retired from research in 1991. He is most famous for his work on light scattering, where he developed the Zimm plot, his extension of the Rouse model of polymer dynamics, and his groundbreaking work on the structure of proteins and DNA. He died on November 26, 2005, at the age of 85 in La Jolla, California.

Figure 40: Portrait of Bruno H. Zimm. Image reprinted with permission from *Macromolecules* **1985**, *18*(11), 2095–2096. Copyright 1985 American Chemical Society.

3.6.4 Relaxation modes

We have learned in Section 2.6.2 that polymers are self-similar and fractal objects. This self-similarity is also valid for chain dynamics: a subchain with g segments in a whole of N segments relaxes like an individual chain composed of just g segments in total, describable by the same Rouse and Zimm formalism as outlined earlier. These subchain relaxations are appraised by so-called **relaxation modes** numerated by an index p. The pth mode corresponds to the coherent motion of subchains with N/p segments in our whole chain with N segments. At $p = 1$, coherent motion of the entire chain is possible, meaning that the entire chain can relax and get displaced by a distance equal to its own size, whereas at $p = 2$, only each half of it can move coherently and relax and therefore get displaced by a distance equal to its own size, respectively. At $p = 3$, just only each third of the chain can move coherently and relax and therefore get displaced by a distance equal to its own size, and so forth. At $p = N$, only single monomeric units can relax and get displaced against each other by their own size. Figure 41 visualizes this hierarchy of relaxation modes. At a time τ_p after abrupt deformation, all modes with index above p are relaxed already, whereas all modes with index below p are still unrelaxed. In general,

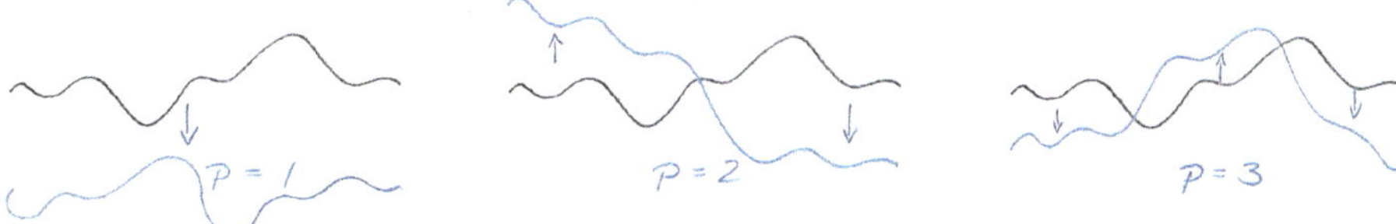

Figure 41: Relaxation modes, indexed by a number p, of a schematic polymer chain. The first mode, $p = 1$, relates to relaxation of the entire chain. In the second mode, $p = 2$, subchain segments with length of just half of the chain can relax. The third mode, $p = 3$, corresponds to the relaxation of subchain segments with length of only a third of the chain, and so on. In the last mode, $p = N$, only single monomeric units can relax (not sketched here). Picture modified from H. G. Elias: *Makromoleküle, Bd. 2: Physikalische Strukturen und Eigenschaften* (6. Ed.), Wiley VCH, **2001**.

the energy storage upon deformation is of order $N \cdot k_BT$. In a deformed viscoelastic material, stress can be relaxed, whereby each mode relaxes a portion of k_BT. This means that the stored energy drops from a total of k_BT per segment at τ_0 to k_BT per chain at τ_{Rouse} or τ_{Zimm}. After an intermediate time of τ_p, only $p \cdot k_BT$ remains stored. From this notion, as well as from an expression for the time dependence of the mode index – an equation that tells us up to which number p modes are already relaxed at a time of interest τ_p – we may derive quantitative expressions for the time-dependent energy storage and relaxation capabilities of polymer materials, which we will do in Section 5.8.2.

Table 8 summarizes the characteristic times that we have determined for the Rouse and the Zimm model as well as the time dependence of the mode index p.

Table 8: Characteristic parameters of the Rouse and Zimm model. This table serves as a toolbox to derive analytical expressions for the time-dependent mechanical spectra of polymer solutions and melts in Chapter 5.

	Rouse model	**Zimm model**
Longest relaxation time	$\tau_1 = \tau_{Rouse} = \tau_0 N^{1+2\nu}$	$\tau_1 = \tau_{Zimm} = \tau_0 N^{3\nu}$
Relaxation time of the pth mode	$\tau_p = \tau_0 \left(\frac{N}{p}\right)^{1+2\nu}$	$\tau_p = \tau_0 \left(\frac{N}{p}\right)^{3\nu}$
Shortest relaxation time	$\tau_N = \tau_0$	$\tau_N = \tau_0$
Time dependence of the mode index[1)]	$p = \left(\frac{\tau_p}{\tau_0}\right)^{\frac{-1}{1+2\nu}} \cdot N$	$p = \left(\frac{\tau_p}{\tau_0}\right)^{\frac{-1}{3\nu}} \cdot N$

[1)] Plugging in a value for τ_p allows us to calculate the mode down to which relaxation has already proceeded after a time of interest τ_p. These equations are obtained from the ones in the second row by simply rearranging those for p.

3.6.5 Subdiffusion

Let us get back to the Einstein–Smoluchowski equation we have introduced in Section 3.6.1. Actually, it is not yet fully complete. To make it generally applicable, we must extend it by an exponent α:

$$\langle(\vec{r}(t)-\vec{r}(0))^2\rangle = 2\ d\ D\ t^{\alpha} \tag{3.35}$$

When $\alpha = 1$, we get the regular Einstein–Smoluchowski equation that describes normal Fickian diffusion. There are, however, many cases in which $\alpha \neq 1$. If $\alpha < 1$, the diffusion is constrained; this situation is called **subdiffusion**. A typical cause for subdiffusion is if there is temporary trapping of the diffusing molecules, which can be the case when they encounter binding sites on their way of motion.[31] By contrast, if $\alpha > 1$, the diffusion is promoted; this situation is called **superdiffusion**. A typical cause for superdiffusion is if the diffusing molecules can ride on the back of carriers for some time and thereby quickly span large distances on their way.[32]

Let us now consider a polymer chain with N segments, and within that, a subchain with N/p segments. This subchain gets displaced by a distance $R = \vec{r}_j(\tau_p) - \vec{r}_j(0)$ of length $l \cdot (N/p)^{\nu}$, which corresponds to its own size, during the time τ_p. Now imagine that we somehow label one monomer segment j on the subchain and follow its displacement. After τ_p it is

$$\text{according to the Rouse model: } \left\langle\left(\vec{r}_j(\tau_p)-\vec{r}_j(0)\right)^2\right\rangle = l^2\left(\frac{N}{p}\right)^{2\nu} = l^2\left(\frac{\tau_p}{\tau_0}\right)^{\frac{2\nu}{1+2\nu}} \tag{3.36a}$$

$$\text{according to the Zimm model: } \left\langle\left(\vec{r}_j(\tau_p)-\vec{r}_j(0)\right)^2\right\rangle = l^2\left(\frac{N}{p}\right)^{2\nu} = l^2\left(\frac{\tau_p}{\tau_0}\right)^{2/3} \tag{3.36b}$$

In the latter two equations, we have first written down a squared variant of the distance $R = \vec{r}_j(\tau_p) - \vec{r}_j(0)$ of length $l \cdot (N/p)^{\nu}$ that we consider here, and then we have substituted the time dependence of the mode index p in the denominator by plugging in the Rouse- or Zimm-model-related expression from Table 8. The resulting exponents for the time dependence of the squared displacements that we obtain with this approach are $\frac{2\nu}{1+2\nu}$ for the Rouse model (this is ½ in an ideal state with

---

31 As an analogy, consider someone on a wine fest in Mainz who has enjoyed the wine too much and therefore conducts a random walk home. The mean-square displacement of that person scales linearly with time according to the Einstein–Smoluchowski equation. If, however, that person is trapped on the way, for example, by meeting other people to chat with (if this is still possible in that state) or by being attracted by more wine stands, the mean-square progress will be less than proportional to time.

32 As an analogy, consider our drunken friend at the wine fest again. This person may accelerate the random-walk home by riding a bus on the way.

$\nu = 1/2$) and 2/3 for the Zimm model, as shown in Figure 42. Both these exponents are smaller than one. Thus, the **segmental motion at times shorter than τ_{Rouse} and τ_{Zimm} is subdiffusive.** This is due to the hindrance imparted on the movement of each monomer segment (such as our labeled one) by its neighbors, to whom it is bound by chemical bonds. Our selected labeled monomer segment can move into a given direction only when the neighboring monomers do so as well. By contrast, when the neighbors do not move into the same direction, our selected monomer is dragged on its motion, thereby forcing its time-dependent displacement to be subdiffusive rather than freely diffusive. Once more, the motion is slower in the Rouse than in the Zimm scenario, as also seen in Figure 42, where the line with slope 1/2 leads to smaller displacements in a given time than the line with slope 2/3. Once more, this is due to the greater viscous drag imparted by the freely draining solvent in the Rouse situation as compared to the solvent drag only on the frontal surface of the solvent-filled coil in the Zimm situation.

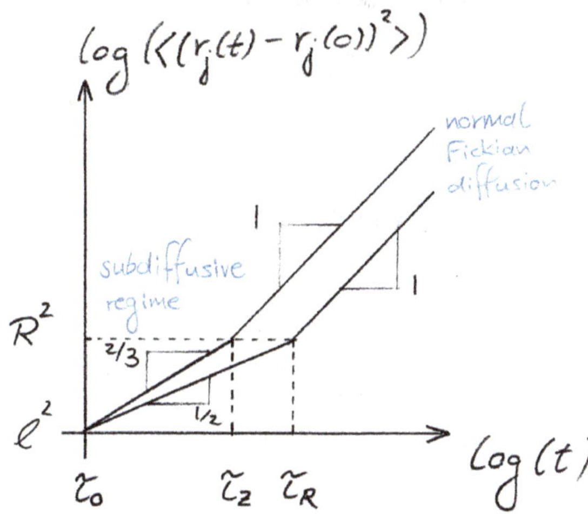

Figure 42: Time dependence of the mean-square displacement of a "labeled" monomer unit (named j) in the Rouse and Zimm scenario. At times smaller than the Rouse- or Zimm-time, the power-law exponent in the two scaling laws is smaller than one, indicating subdiffusion of the labeled segment. Picture redrawn from M. Rubinstein, R. H. Colby: *Polymer Physics*, Oxford University Press, **2003**.

3.6.6 Validity of the models

Now that we have treated two different models that describe the dynamics of a polymer chain, a natural question is which of the two is able to make more accurate predictions. As it turns out, both do, but their validity depends on the polymer's surroundings. In dilute solution, where the polymer concentration is low, hydrodynamic interactions between the segments in a coil are strong. In this case, the Zimm model is the more valid one. We can see evidence for that in the left half of Figure 43, which shows the molar-mass-dependent scaling of the diffusion coefficient of three types of polymers in dilute solution. Each exhibits scaling to the power of the negative Flory exponent that applies to that specific polymer–solvent combination, in perfect agreement to eq. (3.33). By contrast, in the semidilute concentration regime or in a melt, both characterized by marked mutual interpenetration of the coils, the hydrodynamic interactions are screened by overlapping segments of other polymer chains. This effect is similar to the screening of excluded volume interactions in polymer melts that therefore always show θ-type coil conformation. In these regimes, the Rouse model is better suited to describe the polymer chain dynamics.[33] We can see evidence for that in the right half of Figure 43, which shows the molar-mass-dependent viscosity of different polymer melts. In the low molar-mass regime, this quantity rises linearly with the molar mass, as will be proven to be Rouse-model based in Section 5.10.1. (In the high molar-mass regime, the scaling is significantly steeper, indicating a different mechanism of chain motion: the reptation mechanism, which we will also discuss in Section 5.10.1.)

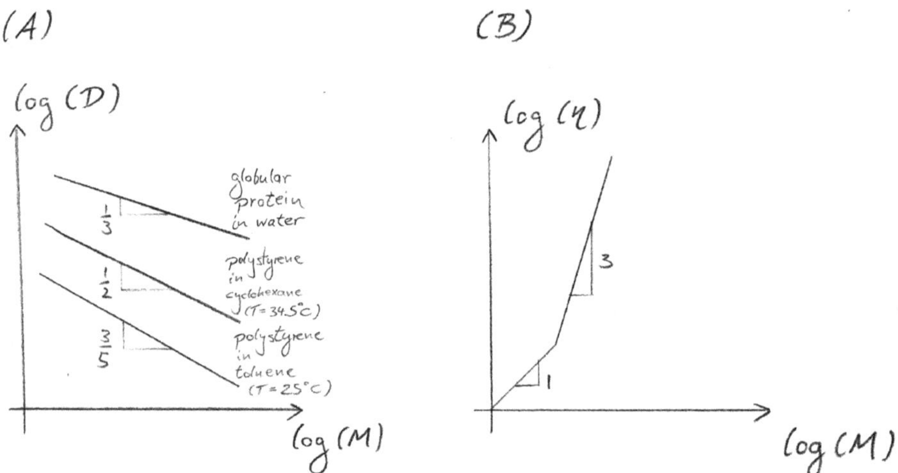

Figure 43: Scaling of (A) the polymer-chain translational diffusion coefficient, D, in dilute solutions and of (B) the polymer-melt viscosity, η, both as a function of the polymer chain length, here assessed by the polymer molar mass, M. Although dilute solutions show scaling of $D \sim M^{-\nu}$, as predicted by the Zimm model, melts of chains with a length shorter than a certain critical one exhibit $\eta \sim M^1$, as predicted by the Rouse model (for a detailed, full explanation of this specific scaling see Section 5.10.1).

33 Actually, in the semidilute regime, polymers display *both* Rouse- and Zimm-type dynamics, depending on the length scale under consideration.

4 Polymer thermodynamics

LESSON 7: FLORY–HUGGINS THEORY

All the previous discussion on polymers in this book was limited to single chains. The following lesson moves beyond and introduces a concept to model and quantify the thermodynamics of mixing of multiple chains with either a low-molecular solvent or with another polymer species. You will see from this modeling that the mixing entropy, which is majorly responsible for the ease of mixing of many substances in classical chemistry, plays a minor role only in polymer science, such that polymers only mix when there is very high enthalpic compatibility. This is captured in a fundamental quantity: the Flory–Huggins interaction parameter.

Throughout this textbook, so far, we have investigated how a polymer chain is shaped, we have learned how it interacts with itself and with a solvent, and we have obtained a picture on how it moves. The common factor in all of this is that, so far, we have focused only on single chains. In the following chapter, we want to expand our view to multi-chain systems. We now ask ourselves: how do many chains interact with one another and with a solvent? Answering that question will be a complex endeavor, because it will require the appraisal of a huge number of interactions, and is, thus, mathematically hard to describe. In addition, the entire multichain system is subject to constant dynamic change, making it even more difficult for us to describe it adequately. So, how do we solve this?

We approach this challenge by just considering *average* monomer–monomer and monomer–solvent interactions. In short, rather than trying to appraise how many M–M and how many M–S interactions we have at a time in the system, and rather than trying to appraise where exactly in the system they occur, we are satisfied with the knowledge of how many of such interactions we have *on average* (over space and time). A very convenient aspect of that simplification is that this number simply scales in proportion to the respective volume fractions of M and S in the system. This approach is called a *mean-field treatment*.

4.1 The Flory–Huggins mean-field theory

Using a mean-field approach, we are able to conceptualize the thermodynamics of polymer–solvent and polymer–polymer mixtures. Said mean-field approach was independently developed by Maurice L. Huggins and Paul J. Flory and is called the

https://doi.org/10.1515/9783110672817-004

Flory–Huggins **mean-field theory**.[34] It appraises the change of the Gibbs free energy of mixing, ΔG_{mix}, which must be negative for two components to readily mix with each other:

$$\Delta G_{mix} = G_{AB} - (G_A + G_B) = \Delta H_{mix} - T\Delta S_{mix} \tag{4.1}$$

Here, G_{AB} is the free energy of the mixture, whereas G_A and G_B are the free energies of the demixed components. ΔG_{mix} can also be expressed in terms of the enthalpy and entropy of mixing, ΔH_{mix} and ΔS_{mix}, respectively. In general, entropy always favors mixing, as that decreases the order of the system, whereas enthalpy usually disfavors mixing, because that would force the components to interact with each other rather than with themselves, which is a less perfect mutual match in most cases.

Maurice Loyal Huggins was born on September 19, 1897, in Berkeley, California. He studied chemistry at the University of California in Berkeley, where he obtained his Master's degree in 1920. He obtained his PhD degree two years later, in 1922, under Charles M. Porter for work on the structure of benzene. He then worked at different institutes, including Stanford Research Institute, Johns Hopkins University (Baltimore, MD), before he joined Eastman Kodak (Rochester, NY) in 1939. From 1959, he returned to Stanford Research Institute and eventually retired from research in 1967. During his active time, he independently conceived the idea of hydrogen bonding in 1919, was an early advocate for its role in stabilizing protein secondary structures, and produced a model of the α-helix in 1943, roughly eight years ahead of the modern model of Linus Pauling, Robert Corey, and Herman Branson. He also developed the mean-field theory for polymer solutions, the Flory–Huggins theory. He died on December 17, 1981, in Woodside, California.

Figure 44: Portrait of Maurice L. Huggins. Image reproduced with permission from Oregon State University Libraries Special Collections & Archives Research Center.

34 Flory himself suggested naming Huggins first (see *J. Chem. Phys.* **1942**, *10*(1), 51–61), as Huggins was in fact the erstwhile of the two scientists to publish the mean-field theory (Huggins' paper was printed in the May issue of *J. Chem. Phys.* **1941**, whereas Flory's paper appeared in the August issue of that journal). By the time of Flory's suggestion, though, the former name was already established in the scientific community, such that it is still used in the order of Flory–Huggins today.

In classical mixtures of low-molar-mass compounds, the inherently favorable entropy influence in eq. (4.1) is powerful enough to outweigh the usually unfavorable enthalpy influence, such that mixing occurs. (And if that does not occur at room temperature, it often helps to heat up, as this will give the entropy term in eq. (4.1) more emphasis.) However, the favorable entropy term is much less powerful for polymer systems. This is because polymers are long chain-like molecules that bring a large number of monomer units into quite some preorder, simply by tying them together in the form of chains. In that state, these monomers cannot arrange themselves in a solvent with that much freedom as they could if they were not polymerized, and as a result, there is much less entropy gain in the polymeric state than it could be in a nonpolymeric state upon mixing with a solvent. Consequently, dissolution and intermixing of polymers is mostly dependent on the enthalpic term, as we have learned already in Section 3.2.

The Flory–Huggins mean-field theory now aims at independently estimating the entropic and the enthalpic contributors to eq. (4.1). It is based on a **lattice model**, as shown in Figure 45. Component A is represented by black beads, whereas component B is represented by white beads. These beads can be placed on lattices, either separate ones for each species (left-hand side of Figure 45), which represents the demixed state, or a common lattice for both species (right-hand side of Figure 45), which represents the mixed state. We assume that all molecules involved, be it the solvent (white beads) or the dissolved component (black beads; note: if that compound is a polymer, the beads are its monomeric units!), have the same volume.

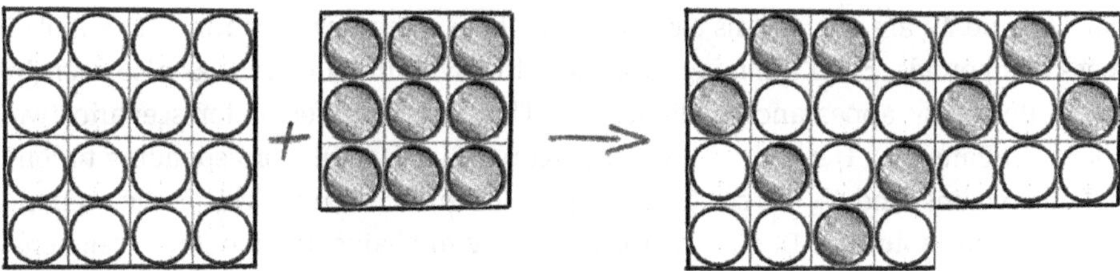

Figure 45: Mixing of black and white beads with the same volume on a lattice. Picture redrawn from M. Rubinstein, R. H. Colby: *Polymer Physics*, Oxford University Press, **2003**.

Based on this premise, the volumes of the individual demixed lattices, V_A and V_B, are additive upon mixing and sum up to the volume of the combined lattice, V_{mix}:

$$V_{mix} = V_A + V_B \tag{4.2}$$

When the mixing is thermodynamically favored (which we aim to appraise here), all beads will randomly occupy new positions within the new combined volume

V_{mix}. The resulting volume fractions[35] occupied by component A and B, ϕ_A and ϕ_B, are given as follows:

$$\phi_A = \frac{V_A}{V_A + V_B} \tag{4.3a}$$

$$\phi_B = \frac{V_B}{V_A + V_B} = 1 - \phi_A \tag{4.3b}$$

We define the unit volume of each single lattice site to be v_0. Then, the total number of lattice sites, n, is

$$n = \frac{V_A + V_B}{v_0} \tag{4.4}$$

Component A occupies $n\phi_A$ of the lattice sites, whereas component B occupies $n\phi_B$ of the lattice sites.

We can also use v_0 to calculate the chain volumes, v_A and v_B, of both components:

$$v_A = N_A v_0 \tag{4.5a}$$

$$v_B = N_B v_0 \tag{4.5b}$$

Here, N_A and N_B are the numbers of *adjacent* lattice sites occupied by each molecule of component A or B, respectively. For a polymeric sample, N_A and N_B correspond to the degrees of polymerization of the polymer chains (which scales to their sizes as $R_A \sim N_A^{\nu}$ and $R_B \sim N_B^{\nu}$).

We can delimit three different cases, depending on the relative size of N_A and N_B, all three of which are shown in Figure 46. In the first scenario, depicted in Panel (A), both N_A and N_B are one. In this case, we talk about a **regular solution** of low-molar-mass compounds. Such a mixing process is highly entropy driven, as also seen by the quite messy appearance of Panel (A). This is very different for scenario two, shown in Panel (B). Here, $N_A >> 1$ (in the figure, it is $N_A = 10$), corresponding to long polymer chains, and $N_B = 1$, corresponding to solvent molecules. This is the scenario of a **polymer solution.** Due to their connectivity to chains, the monomer segments now have much less possibilities to arrange themselves than in the first scenario, as seen by the more tidy distribution of the black beads in Panel (B) as compared to Panel (A); only the solvent molecules still can arrange themselves quite freely in that kind of system. In scenario three, shown in Panel (C), both N_A and N_B are $>> 1$ (in the figure, $N_A = N_B = 10$). This is the case for a mixture of two different polymers, a

35 The volume fraction is a common measure of concentration in polymer science. It denotes what fraction of the volume of a system under consideration is occupied by a species of interest. You know a related quantity from elementary physical chemistry: the mole fraction. If there's two components A and B, it calculates as $x_A = \frac{n_A}{n_A + n_B}$ and $x_B = \frac{n_B}{n_A + n_B} = 1 - x_A$.

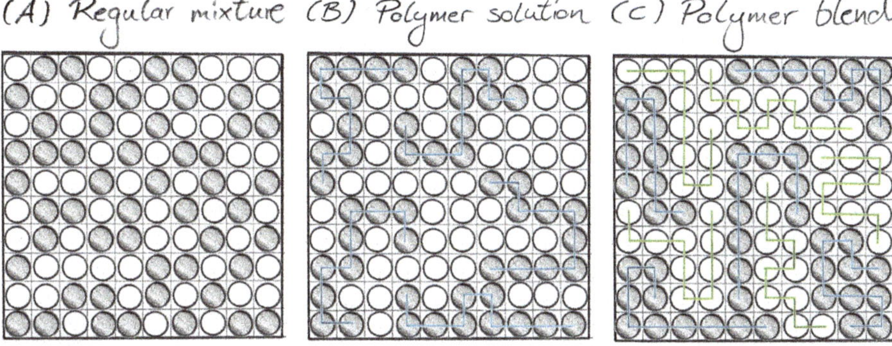

Figure 46: Arrangement of 50 black and 50 white beads on a 10 × 10 lattice with elementary-unit volume v_0 to model the binary mixing of two components in **(A)** a regular solution of two low-molar-mass components A and B, both consisting of molecules with size equal to the lattice-site size, **(B)** a polymer solution of a low-molar-mass component A that is built of monomer units (here: 10 per chain) that each occupy one lattice site, such that each macromolecule of the whole species occupies N_A adjacent sites, dissolved in a low-molar-mass solvent B that consists of molecules equal to the lattice-site size, and **(C)** a polymer blend of one high-molar-mass species A mixed with another high-molar-mass species B, both occupying multiple adjacent lattice sites, N_A and N_B (here: both 10 per chain). Pictures redrawn from M. Rubinstein, R. H. Colby: *Polymer Physics*, Oxford University Press, **2003**.

polymer blend. In that case, the freedom of arrangement of both the black and the white beads is quite limited, as both are constrained by their connectivity to chains.

4.1.1 The entropy of mixing

The entropy of mixing, ΔS_{mix}, can be estimated by the number of microstates that realize the two macrostates *mixed* and *demixed*. This is a statistical approach, through which we can calculate the entropy by Boltzmann's formula, $S = k_B \ln(W)$. In a homogeneous A–B mixture, each bead has n possible positions where it can be placed on the common lattice. In other words, each bead can sit anywhere on the lattice. In the demixed phases, component A only has $n\phi_A$ possible positions in the A phase; the rest is occupied by component B in the B phase. From this, we get the entropy of mixing for a single A-bead, ΔS_A, as[36]

$$\Delta S_A = k_B \ln n - k_B \ln n\phi_A = k_B \ln \frac{1}{\phi_A} = -k_B \ln \phi_A \tag{4.6a}$$

36 Do not confuse the index B at k_B, where it denotes *Boltzmann*, to the index B at n_B, N_B, and ϕ_B, where it denotes *species* B.

Here, $k_B \ln n$ relates to the mixed, and $k_B \ln n\phi_A$ to the demixed state.

Analogously, the entropy of mixing for a single B-bead is

$$\Delta S_B = -k_B \ln \phi_B \tag{4.6b}$$

With both these expression, we can calculate the total entropy of mixing, ΔS_{mix}

$$\Delta S_{mix} = n_A \Delta S_A + n_B \Delta S_B = -k_B(n_A \ln \phi_A + n_B \ln \phi_B) \tag{4.7}$$

If we express the number of A-molecules, which are single beads in the case of a low molar-mass substance or chains built up of multiple beads in the case of a polymer, n_A, as $n \cdot \phi_A / N_A$ as well as the number of B-molecules, n_B, as $n \cdot \phi_B / N_B$ in eq. (4.7) and then divide by n, we get the entropy of mixing per lattice site:

$$\Delta \bar{S}_{mix} = \frac{\Delta S_{mix}}{n} = -k_B \left(\frac{\phi_A}{N_A} \ln \phi_A + \frac{\phi_B}{N_B} \ln \phi_B \right) \tag{4.8}$$

In this equation, note again that ϕ_A and ϕ_B are the volume fractions of the components A and B, and N_A and N_B are their degrees of polymerization, meaning how many adjacent lattice sites are occupied per A and B (macro)molecule. If we apply this formula to a regular mixture, with $N_A = N_B = 1$, and with equal molar volumes of the A- and B-beads (which we assume allover the present discussion anyways), we may replace the volume fractions in eq. (4.8) by mole fractions, thereby yielding a form of the equation that you know from elementary physical chemistry: $\Delta \bar{S}_{mix} = -k_B(x_A \ln x_A + x_B \ln x_B)$.

When inspecting eq. (4.8), we can get some conceptual insights. As the numerical values for ϕ_A and ϕ_B are both 0…1, the logarithms in eq. (4.8) will always be negative (or zero at most). Together with the minus sign in front of the right-hand side of the equation, this will always lead to a positive mixing entropy. That makes sense, as entropy always favors mixing. The extent of positivity, however, depends on the denominators N_A and N_B. The larger they are, the less positive will be the mixing entropy. This means that, even though entropy always favors mixing, it does so with less power at high chain length. This insight matches the qualitative discussion that we have just led on the preceding pages.

Let us deepen our conceptual insight by some quantitative calculation and estimate the entropy of mixing for each of the three cases displayed in Figure 46. Here, the volume fractions of component A and B are equal, $\phi_A = \phi_B = 0.5$. In our first scenario, Panel (A), we have 50 white and 50 black beads that we can arrange freely on the lattice. This means that there are $100! / (50!\ 50!) = 10^{29}$ different possible microstates to realize the macrostate "mixed". The entropy of mixing per lattice site normalized by Boltzmann's constant, $\Delta \bar{S}_{mix}/k_B$, is 0.69 in this case. In Panel (B), our second scenario, we retain the 50 white solvent beads, but change the structure of component A, represented by the black beads, such that it is now a 10-segment polymer, thereby now represented by five 10-segment chains of black beads. Because the black beads are connected to each other in this fashion, the normalized entropy of mixing is markedly reduced, $\Delta\bar{S}_{mix}/k_B = 0.38$. In the last scenario, Panel (C), this trend continues. Now, the white beads are also connected together in five polymer chains with 10 segments each. As a result, the number of possible arrangements on the lattice for both components drops to only about 10^3. Consequently, the normalized entropy of mixing is further reduced, $\Delta\bar{S}_{mix}/k_B = 0.069$.

To visualize these findings, the normalized entropy of mixing is plotted as a function of the volume fraction of component A in Figure 47. Again, we realize from this plot that the entropic gain from mixing a polymer with a solvent is drastically lower than that of a regular mixture of low-molar-mass compounds. Furthermore, we see that the entropic gain of the polymer blend is almost negligible. What we also realize from this plot is that the extent of the entropic gain also depends on the volume fraction; it is greatest at a 50:50 mixture ($\phi_A = 0.5$) if the mixing components have equal sizes, which is the case for our scenario A with $N_A = N_B = 1$ and for our scenario C with $N_A = N_B = 10$, whereas the entropy maximum is at a different ϕ_A (of about 0.65) if the mixture consists of compounds with different sizes, as it is the case in scenario B, where we have $N_A = 10$ but $N_B = 1$.

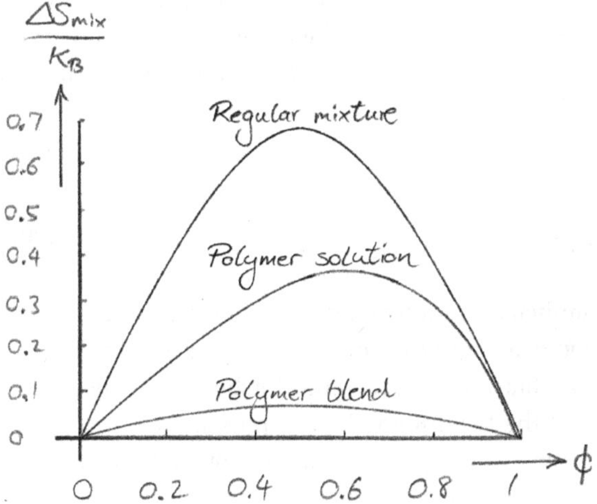

Figure 47: Entropy of mixing for the three different cases shown in Figure 46.

The just minor gain of entropy upon mixing polymers with solvents or other polymers is one of two reasons for nature to rely the build-up of complex structures on polymeric building blocks: since their mess-up would bring just minor entropic gain, their tiding to ordered structures does not cost much either.[37]

37 The second reason is that, due to their interaction energies of just a few couple *RT*, polymeric building blocks can serve to build entities that are stable enough to persist, but also labile enough to be de- and rearrangeable with not too much energy input.

4.1.2 The enthalpy of mixing

Let us now focus on the energetic contributions to the mixing of two components A and B. To do this, we have to quantify all possible interactions, which are A–A, B–B, and A–B. This, however, cannot be done analytically. Remember that a regular 1:1 mixture of even just 50 A-beads and 50 B-beads gives 10^{29} possible combinations. To address this challenge, we use a mean-field approach. In that approach, we conceptually "hydrolyze" our polymer to unconnected segments that have the same size as their surrounding molecules, and distribute them randomly on the lattice. We then only discuss interactions between two direct neighbors on this lattice. This gives us three different interaction energies, u, that we must discuss: u_{AA}, u_{BB}, and u_{AB}.

The **average pairwise interaction energy** of a monomer A with one of its neighboring lattice sites is given as

$$U_A = u_{AA}\phi_A + u_{AB}\phi_B \tag{4.9a}$$

Here, u_{AA} is the interaction energy in case of an A–A contact and ϕ_A the probability for the neighboring site of the A-monomer under consideration to be indeed occupied by another A. This value is equal to the volume fraction of A in the system. The second term denotes the same for the case of an A–B contact.

Analogously, the average pairwise interaction energy of component B with one of its lattice sites is

$$U_B = u_{AB}\phi_A + u_{BB}\phi_B \tag{4.9b}$$

Arguing this way exposes the simplicity and advantage of a mean-field approach. Normally, the interaction of one molecule (or bead) of component A is either *exactly* u_{AA} or *exactly* u_{AB}, depending on what type of molecule (or bead) sits next to it. However, we do not know which of the two interactions is present at a given time and spot on the lattice, so we assume an *average mean interaction* of both based on their frequency of occurrence in the system, ϕ_A and ϕ_B.

We know that each of the n lattice sites has z nearest neighbors to interact with. From the single component average pairwise interaction energies, we can calculate the total average interaction energy *in the mixed state*:

$$U = \frac{z \cdot n}{2}(U_A\phi_A + U_B\phi_B) \tag{4.10a}$$

with U_A and U_B given by eq. (4.9a) and (4.9b). The factor of ½ rows back double-counting of the pairwise interactions.

The average interaction energy *in the demixed states* can also be determined. In this case, we use the individual interaction energies, u_{AA} and u_{BB}, in an expression just like eq. (4.10a):

$$U_0 = \frac{z \cdot n}{2}(u_{AA}\phi_A + u_{BB}\phi_B) \tag{4.10b}$$

With the interaction energies in the mixed and demixed states, we can calculate the energy change upon mixing. We do this by subtracting the energies and normalize the resulting difference by dividing it by the number of lattice sites, n:

$$\Delta\bar{U}_{mix} = \frac{U - U_0}{n} = \frac{z}{2}(2u_{AB} - u_{AA} - u_{BB})\,\phi_A\phi_B = k_B T\chi\phi_A\phi_B \tag{4.11}$$

This equation introduces a very important quantity for the discussion of polymer thermodynamics: the Flory–Huggins **interaction parameter**:

$$\chi = \frac{z}{2} \cdot \frac{2u_{AB} - u_{AA} - u_{BB}}{k_B T} \tag{4.12}$$

This parameter is a dimensionless measure of the difference of the pairwise interaction energies before mixing (u_{AA} and u_{BB}) and after mixing ($2u_{AB}$), normalized to the lattice geometry ($z/2$) and the elementary thermal-energy increment ($k_B T$). During mixing of A and B, we break A–A and B–B contacts but in turn establish two new A–B contacts per such breaking. This means that we "loose" u_{AA} and u_{BB}, but we gain $2u_{AB}$. Hence, according to eq. (4.12), when $\chi < 0$, this means that there is more energy gained than lost upon this exchange, as in that case, $2u_{AB}$ is more negative (= thermodynamically more favorable) than u_{AA} and u_{BB} are together; this corresponds to an **exothermic mixing** process. When $\chi > 0$, by contrast, there is more energy lost than gained upon mixing, as in that case, $2u_{AB}$ is less negative (= thermodynamically less favorable) than u_{AA} and u_{BB} together; this corresponds to an **endothermic mixing** process. A Flory–Huggins interaction parameter of $\chi = 0$ indicates no energy change upon mixing. The mixing process is then driven by its entropic contribution only, a situation commonly referred to as **ideal mixture**. Most usually, mixing of two components A and B is disfavorable, as two A–B contacts just cannot be as perfect matches as an A–A and a B–B contact. This corresponds to an endothermic mixing process, characterized by a positive χ.

If we assume, for the sake of simplicity, that our mixing process is isochoric, which means that the total volume does not change upon mixing, $\Delta V_{mix} = 0$, then we can substitute the enthalpic term, ΔH_{mix}, in eq. (4.1) by ΔU_{mix}, that is, by eq. (4.12). This yields the change of the Gibbs free energy upon mixing:

$$\Delta\bar{G}_{mix} = k_B T\left(\frac{\phi_A}{N_A}\ln\phi_A + \frac{\phi_B}{N_B}\ln\phi_B + \chi\phi_A\phi_B\right) \tag{4.13}$$

We have already stated that ΔG_{mix} must be negative for mixing to occur, so let us look a little bit closer at the latter equation. The combinatorial entropy term always contributes something negative, as the numerical values for ϕ_A and ϕ_B are both 0 … 1, so the logarithms in eq. (4.13) will always be negative (or zero at most). This makes sense, as entropy always favors mixing. For the enthalpic term, however, it depends on the difference of the pairwise interaction energies before and after mixing, indicated by the Flory–Huggins parameter χ. As just said above, in most cases, $\chi > 0$, because the components rather want to interact with themselves than with one another. The enthalpic term, then, counteracts the entropic contribution. It therefore depends on how strong the entropic contribution is, and how large the counteracting enthalpic term is, which is most dominantly regulated by the magnitude of χ. As we have discussed earlier in this chapter, the entropy of mixing is very small for polymer solutions and blends (see Figure 47). As a result, χ must be lower than a critical value (which is 0.5 in the case of polymer solutions and close to zero in the case of polymer blends, as will be demonstrated in Section 4.2) for mixing still to happen. Just in very rare cases, χ can be smaller than 0. For that to be, there must be a mutually complementary favorable interaction between A and B, such that there will be an additional energy gain when both components interact with each other rather than with themselves. A clever realization of that premise was presented in Freiburg and Mainz in the 1980s by Reimund Stadler. He equipped polymer chains with complementary hydrogen bonding motifs, as shown in Figure 48. The energy gained from these transient interactions very much fuels the mixing of these polymers.

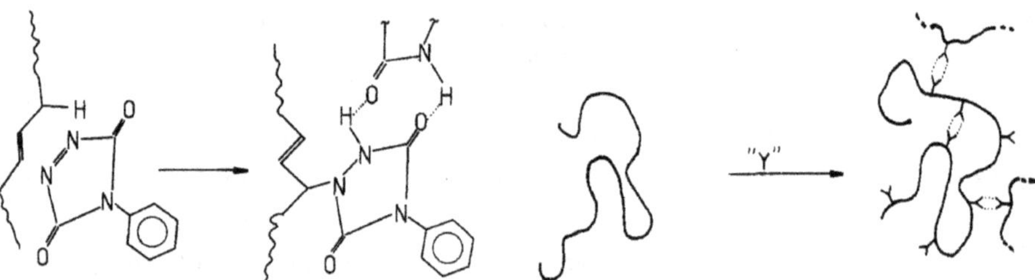

Figure 48: Reimund Stadler's approach of equipping polymers with self-complementary hydrogen bonding motifs to fuel a favorable enthalpic contribution to the free energy of mixing. Pictures reprinted with permission from R. Stadler, L. L. de Lucca Freitas, *Coll. Polym. Sci.* **1986**, *264*(9), 773–778, copyright 1986 Steinkopff Verlag (now Springer Nature), and R. Stadler, L. L. de Lucca Freitas, *Macromolecules* **1987**, *20*(10), 2478–2485, copyright 1987 American Chemical Society.

Reimund Stadler was born on October 9, 1956, in Stühlingen, Germany. He studied chemistry at the Albert Ludwigs University Freiburg, where he received his diploma and PhD degree, with a thesis on "Viscoelasticity and Crystal Melting of Thermoplastic Elastomers". He then became a postdoc at Porto Alegre in Brazil before his habilitation in 1989 under Hans-Joachim Cantow back in Freiburg. Stadler was appointed as full professor at the Johannes Gutenberg University Mainz, and then switched to the University of Bayreuth, where he died suddenly just one year later, on June 14, 1998. To his honor, the German Chemical Society gives out an award named after him to rising-star researchers in the field of polymer science every other year.

Figure 49: Portrait of Reimund Stadler. Image reproduced with permission from *Designed Monomers and Polymers* **1999**, 2(2), 109–110. Copyright 1999 Taylor & Francis.

4.1.3 The Flory–Huggins parameter as a function

The Flory–Huggins theory is a simplified approach to polymer thermodynamics: it appraises the entropy of mixing based on a combinatorial argument, and the energy of mixing based on mutual average interactions in a mean-field treatment. So far, so good. There are, however, several aspects that are not at all or at least not adequately captured by that theory, for example, additional entropic effects such as the need for favorable mutual orientation of the molecules to each other to interact, or mistakes in the enthalpic part of the theory that come from the mean-field treatment, ignoring the rather inhomogeneous distribution of the two species in the system, where we have high densities of the black species inside the polymer coils, but none of it between the coils. All these deviations from reality are lumped into the Flory–Huggins parameter χ. It is therefore not a given and fix number, but a function that itself is composed of an entropic part, χ_S, and a temperature-weighted enthalpic part, χ_H.

$$\chi = \chi_S + \frac{\chi_H}{T} \tag{4.14a}$$

It is therefore the interplay of both contributions, χ_S and χ_H, and temperature that determines the miscibility or immiscibility of a polymer and a solvent. The role of the entropy part and the enthalpy part of χ can be understood based on the following thought: Let us consider *all* contributions to the mixing free enthalpy, that means, not only the simple configurational entropy term and the simple pairwise-interaction enthalpy term in the Flory–Huggins eq. (4.13), but also all further specific entropic and enthalpic effects that the mean-field approach doesn't account for; we name this full set of contributions $\Delta G_{\text{mix, excess}}$. We may then write the following equation to connect this to the (then phenomenological) parameter χ:

$$\Delta G_{\text{mix, excess}} = \Delta H_{\text{mix, excess}} - T \cdot \Delta S_{\text{mix, excess}} = k_B T \cdot \left(\chi_S + \frac{\chi_H}{T}\right) \tag{4.14b}$$

Based on that, we identify $-k_B \cdot \chi_S$ as an excess mixing entropy $\Delta S_{\text{mix, excess}}$ that is not based on simple statistical-combinatorial considerations, but lumps all the specificity of the molecules acting as the solvent and the solute polymer. $k_B \cdot \chi_H = \Delta H_{\text{mix, excess}}$ still reflects the mixing enthalpy based on mean-field average pairwise interactions in the form of $\chi_H = (z/2k_B)(2u_{AB} - u_{AA} - u_{BB})$, so that $\Delta H_{\text{mix, excess}} = (z/2)(2u_{AB} - u_{AA} - u_{BB})$. (Note that in this notion, χ_H has a physical unit [K], whereas χ_S has no unit.)

Let us examine the enthalpic contribution to the Flory–Huggins parameter, χ_H, a little more closely. In unpolar systems, it is commonly positive, $\chi_H > 0$. This is due to mostly attractive *intra*molecular interactions such as van der Waals interactions in those systems, which oppose mixing as they are better built between each component itself but not so perfectly between the components each other.[38] These interactions, however, can be overcome at high temperatures, such that mixing is promoted at high temperatures. Mathematically, this situation is reflected by the temperature dependence in the denominator of eq. (4.14a). If $\chi_H > 0$, then we add something positive to χ_S, such that our total χ gets larger; this is bad for mixing, as χ should be small for that, smaller than a certain critical limit (which we will discuss in the next section). As T reduces the χ_H contribution by its position in the denominator, however, this unfavorable part of χ is attenuated if T is high. In the opposite case, usually encountered in polar, protic systems, χ_H can be negative, that is, $\chi_H < 0$. This situation is encountered if we have specific *inter*component interactions, such as hydrogen bonds or dipole–dipole interactions. These transient bonds favor the mixing, but they are broken at high temperature. Mathematically, we can see this again from the temperature dependence in the denominator of eq. (4.14a). If $\chi_H < 0$, then we add something negative to χ_S, such that our total χ gets smaller; this is good for mixing. As T reduces the χ_H contribution by its position in the denominator, however, this favorable part of χ is attenuated at high T.

A prime example of the latter case is an aqueous solution of poly(*N*-isopropylacrylamide), pNIPAAm, whose structure is depicted in Figure 50. This polymer has a largely unpolar backbone and an unpolar side group that both disfavor interactions with water, but it also has a polar amide group that can form hydrogen bonds with water molecules. This promotes mixing overall, but only up to a temperature of 32 °C. At higher temperatures, the hydrogen bonds break and the polymer precipitates,

38 A quantitative justification for $\chi_H > 0$ in systems that prefer intra- over intercomponent contact is given in Footnote No. 45 in Section 4.2.1.

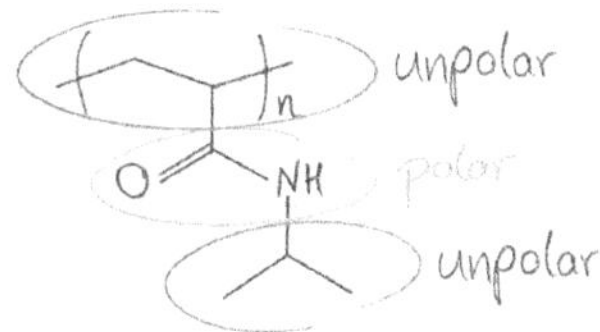

Figure 50: Poly(*N*-isopropylacrylamide) is a polymer with a largely unpolar backbone and an unpolar side group that both dislike water, whereas a polar amide group in between can promote mixing with water due to hydrogen bonding. The hydrogen bonds, however, are broken at high temperature, and the polymer then precipitates from the solution.

because then, the so-called hydrophobic effect comes into play: Water molecules that find themselves in close proximity to hydrophobic domains, such as the isopropyl moieties in pNIPAAm, form clusters and therefore assume a more ordered intermolecular arrangement than they would have in their natural state, because in that ordered arrangement they avoid unfavorable hydrophobic–hydrophilic contact. This effect causes an additional unfavorable excess mixing entropy $\Delta S_{\text{mix, excess}} < 0$. With increasing temperature, we find that the contribution of this excess entropy to $\Delta G_{\text{mix, excess}}$, which is $-T \cdot \Delta S_{\text{mix, excess}}$, increases linearly, and therefore the excess free enthalpy of mixing $\Delta G_{\text{mix, excess}}$ can compensate the usual configurational Flory–Huggins free enthalpy of mixing, eventually leading to $\Delta G_{\text{mix, total}} > 0$, thereby inducing phase separation.

Such a solubility-temperature barrier is called **Lower Critical Solution Temperature**, LCST. There is a whole class of LCST-type polymers such as pNIPAAm, all of which exhibiting LCSTs reasonably close to the human body temperature.[39] This makes these polymers interesting for biomedical applications, possibly in targeted drug-release systems. For example, consider that inflammatory tissue has a little higher temperature than healthy tissue. If an LCST polymer is designed such to display a coil-to-globule (i.e., swollen-to-deswollen) transition right in the range of that temperature difference, a nanocapsule system may be made from that polymer such that it can collapse and thereby release an anti-inflammatory drug only in the inflammatory tissue regions, but not in the healthy ones. Furthermore, we may expect that protonation or deprotonation of these polymers may further drastically affect their LCST, and with that, their regions of solubility and nonsolubility in water. As a result, we may also use these polymers for pH-dependent active systems, such as nanocapsules that release drugs only in cancer tissue but not in healthy tissue, making use of the circumstance that there is a pH difference between these two kinds of tissues.

As a closing remark, note that in addition to the temperature dependence, the Flory–Huggins parameter also often exhibits a concentration dependence that may be captured in a virial-series form:

---

39 This is because hydrogen-bonding interactions have a strength of some few singles to tens of RT, which is easily activated to open in the temperature window of 30 … 50 °C. Nature uses the same principle for immune reactions (e.g., for fever, which denaturates hostile proteins by breaking their hydrogen-bonding interactions) or to bind and unbind functional entities on demand, such as the double strands of DNA.

$$\chi = \chi_S + \frac{\chi_H}{T} + \chi_1 \phi + \chi_2 \phi^2 + \cdots \tag{4.15}$$

This does even more underline our above statement that χ is not a unique number, but a phenomenological parameter that itself depends on multiple influences, the most relevant being temperature (variable T in eq. (4.15)), composition of the system (variable ϕ in eq. (4.15)), and chain length (not included in eq. (4.15); there could be further terms in it to account for that variable as well).

4.1.4 Microscopic demixing

We have seen that the entropic gain of a polymer–polymer blend is negligible. As a consequence, it is almost impossible to dissolve one polymer in another. Even when the polymers are almost chemically identical, the enthalpic penalty is still high enough to cause demixing.[40] This immiscibility has severe consequences for systems where two polymers are chemically connected to one another, as it is the case in block copolymers. In these systems, the two blocks usually want to demix, but they can do so only on nanoscopic scales, because they are tied together. As a result, the block-copolymer system will show **microphase separation**, with a morphology that depends on the block lengths, as shown in Figure 51. At equal block

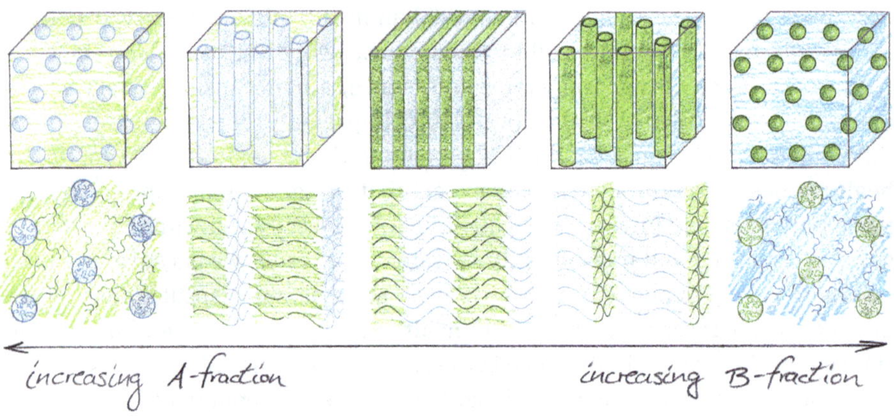

Figure 51: Various block copolymer phases that can form depending on the fraction of the constituent blocks A and B.

40 Polystyrene and deuterated polystyrene are a prime example of this. Both are chemically almost the same (they just differ by exchange of hydrogen for deuterium), but they still have a positive, albeit very small, Flory–Huggins parameter, $\chi = 10^{-4}$. This causes both polymers to be immiscible if their molar mass is greater than about $M = 3 \times 10^6$ g·mol^{-1}.

length, shown in the center part of Figure 51, both blocks will arrange themselves in lamellar phases, whereas if one of the blocks is shorter, it will undergo coiling while the segments of the longer block will be less coiled, thereby changing the lamellae thickness. If the block-length mismatch gets too severe, different other morphologies will result, as shown in the left and right outer parts of Figure 51. These have astonishing order on both microscopic and mesoscopic scales, which is determined by the mutual pinning of locally microseparated phases by chain-blocks that partition in each of them. The overall phase microstructure therefore results from the length, degree of coiling, and extent of partition of either chain block. The astonishing order of the different morphologies shown in Figure 51 is therefore not due to spontaneous self-assembly of the polymers to such states, but instead, due to the hindered *dis*-assembly of the immiscible polymer blocks that results from their connectivity.

Note that a stark contrast to block copolymers is random or alternating copolymers. These copolymers manage to incorporate two chemically different building-block species along their chains. As such, they can serve to incorporate both these components in one polymeric system, without chain–chain immiscibility issues. This circumstance makes random or alternating copolymerization so attractive, as it is often the only way to realize such a material combination, due to the foresaid extreme difficulty of realizing polymer blends. So to speak, in a random or alternating copolymer, the mixing of the different chemical species is done "along the chain backbone".

4.1.5 Solubility parameters

The Flory–Huggins parameter quantifies polymer–solvent interactions; it does so, however, for *pairs* of these only, such that it is somewhat unhandy to tabulate it. More practical would be pairs of parameters, one for the polymer and one for the solvent, that can be tabulated independently and then taken together to appraise the miscibility of both compounds. Such parameters have been introduced by Hildebrandt and Scott. They are based on quantifying the attractive interactions of the solvent molecules or the monomer units of a polymer chain amongst themselves via their **cohesive energy**, ΔE_{A}, which can be determined by measurement of the heat of combustion. With that, Hildebrandt and Scott defined

$$\delta_{\mathrm{A}} = \sqrt{\frac{\Delta E_{\mathrm{A}}}{v_{\mathrm{A}}}} \tag{4.16}$$

Here, δ_{A} is the solubility parameter for compound A, ΔE_{A} is its cohesive energy, and v_{A} its molecular volume.

In the lattice model that we have discussed above, the interaction energy for a given lattice site in a plain-A system is $\frac{z}{2}u_{AA}$, which is equal to $-v_0\frac{\Delta E_A}{v_A}$ and thus, according to eq. (4.16), also equal to $= -v_0{\delta_A}^2$:

$$\frac{z}{2}u_{AA} = -v_0\frac{\Delta E_A}{v_A} = -v_0{\delta_A}^2 \tag{4.17a}$$

The interaction energy of the second component in a plain-B system is calculated analogously

$$\frac{z}{2}u_{BB} = -v_0\frac{\Delta E_B}{v_B} = -v_0{\delta_B}^2 \tag{4.17b}$$

By the geometric mean of both parameters, δ_A and δ_B, we can estimate the mutual interaction energy, u_{AB}

$$\frac{z}{2}u_{AB} = -v_0\delta_A\delta_B \tag{4.17c}$$

Strictly speaking, the solubility parameter as defined above exists only for liquid substances, that is, only for solvents, but not for polymers, as the latter do not exhibit combustion. It can be presumed, though, that this is only due to the covalent connectivity of the monomers, such that if this connection would be lost, they would combust just as an own independent substance composed of molecules with a structure that matches the one of the repeating units in the polymer. Thus, the solubility parameter of the polymer matches the one of such a hypothetic fluid.[41] In other words, although the solubility parameter of a solvent can be measured directly from its heat of combustion, the one for a polymer can be appraised by finding a fluid that has a best identical structure to the monomer units in the polymer and then estimate the solubility parameter of that fluid. As a practical alternative, we may also just seek a fluid that best dissolves the polymer, as assessed by maximum coil expansion in that solvent, which we may determine by light scattering or viscometry, and then presume this polymer–solvent mixture to be an athermal one with interaction energies of kind $u_{AA} = u_{BB} = u_{AB}$, such that we can set the solubility parameter of the polymer practically equal to that of the solvent.

The Flory–Huggins parameter, χ, is related to the solubility parameters by

$$\chi = \frac{v_0}{k_B T}(\delta_A - \delta_B)^2 \tag{4.18}$$

41 Based on our classification in Section 3.2, that fluid would be an athermal solvent for the polymer.

With that equation, χ can be directly calculated from the δ-parameters of a polymer and solvent of interest. Solubility parameters therefore have the advantage that they can be used singularly for these two components and do not have to be specified for each possible pair, as it is the case with the Flory–Huggins parameter. In this way, lists such as Table 9 can be compiled that collect many different solubility parameters, and these collections can then be used to select a good solvent for a given polymer. For this purpose, we just have to choose a solvent with a solubility parameter that is similar to that of the polymer. This even works for solvent mixtures, because the net-δ value is the average of its constituents. In that way, solvent mixtures tailored to the polymer's solubility parameter can easily be prepared.[42]

We see that χ can only be positive according to eq. (4.18). That implies perfect solubility of the two components when $\delta_A = \delta_B$, and thus $\chi = 0$, according to the proverb "*similia similibus solvuntur*", meaning "like dissolves like". Note that δ-values are large for polar solvents, because these have high cohesive energies ΔE_A. Such polar solvents actually require a three-dimensional solubility parameter that takes into account possible secondary interactions:

$$\delta_{3d} = \left(\delta_{\text{vdW}}{}^2 + \delta_{\text{Dipole-Dipole}}{}^2 + \delta_{\text{H-Bonds}}{}^2\right)^{1/2} \tag{4.19}$$

Table 9: Solubility parameters according to Hildebrandt and Scott for some typical solvents and polymers.

Solvent	δ (cal·cm^{-3})$^{1/2}$	Polymer	δ (cal·cm^{-3})$^{1/2}$
Cyclohexane	8.2	Polyethylene	7.9
Benzene	9.2	Poly(vinyl chloride)	8.9
Chloroform	9.3	Polystyrene	9.1
Acetone	9.9	Poly(methyl methacrylate)	9.5
Methanol	14.5	Polyamine 66	13.6
Water	23.4	Polyacrylonitrile	15.4

The unit for the solubility parameter, (cal·cm^{-3})$^{1/2}$, is commonly named 1 Hildebrandt.

42 Most interestingly, even mixtures of two *non*solvents can constitute a solvent for a polymer with a δ in the middle range of Table 9, namely if the mixture brings together one nonsolvent with a too large δ and another nonsolvent with a too small δ, whose average will then lie in the middle and thereby match the δ of the polymer.

4.2 Phase diagrams

LESSON 8: PHASE DIAGRAMS

Based upon the Flory–Huggins theory introduced in the last lesson, the following will show at what combinations of relevant variables, mostly temperature and composition, a polymer will or will not mix with a solvent or with another polymer. A plot of these parameters gives a phase diagram, of which the following lesson will introduce two fundamentally different types for polymer systems.

The Flory–Huggins theory has given us conceptual and quantitative insight about the miscibility of polymers with solvents or with other polymers. In the following, we want to expand this insight to make even more detailed quantitative predictions at what conditions polymer–polymer and polymer–solvent systems are miscible or immiscible. In other words, we want to construct **phase diagrams**.

A phase diagram is a *plot of two practically relevant variables* against each other such to create a map with regions that show at what combination of these variables the system is in what state. You may remember the most simple case of a phase diagram from your elementary physical chemistry classes: a one-component phase diagram, which is a map of *pressure versus temperature* that delimits at which p–T pair the substance is a solid, a liquid, or a gas. This map also delimits in which regions we have coexistence of two or even three of these phases. In general, the construction of a phase diagram is based on seeking the minimum of the free energy for each region in the map. In the following, we want to derive such a map for two-component polymer mixtures; thus, one of our two practically relevant variables will be the system's composition, that is, the volume fraction of the polymer of interest in the mixture, ϕ. The other variable will be a measure of the energetic interactions between the components. Naturally, this variable is χ but as we will see later, that can actually be translated easily to an even more practical variable, T.

4.2.1 Equilibrium and stability

The mixing and demixing of a system with a component A at the initial composition ϕ_0 is determined by the free energy of its mixed phase, $F_{mix}(\phi_0)$, and the free energy of the separated phases α and β, $F_{\alpha\beta}(\phi_0)$. When $F_{mix}(\phi_0) < F_{\alpha\beta}(\phi_0)$, then the mixture is stable; in this case, when brought together, both components will mix spontaneously and stay mixed. By contrast, when $F_{mix}(\phi_0) > F_{\alpha\beta}(\phi_0)$, then the mixture is unstable and will phase-separate, or the components would not even mix in the first place. To determine this ratio, we need to determine functions that describe $F_{\text{mix}}(\phi_0)$ and $F_{\alpha\beta}(\phi_0)$.

In the phase-separated state, an initial system composition ϕ_0 is constituted by

$$\phi_0 = f_\alpha \phi_\alpha + f_\beta \phi_\beta \tag{4.20}$$

Here, f_α is the relative fraction of phase α, and ϕ_α is the volume fraction of component A in that phase α. Analogously, f_β is the relative fraction of phase β, and ϕ_β is the volume fraction of component A in that phase β.

The energy of the phase-separated state, $F_{\alpha\beta}(\phi_0)$, is a simple sum of similar kind: $F_{\alpha\beta} = f_\alpha F_\alpha + f_\beta F_\beta$. By combination with eq. (4.20), we get

$$F_{\alpha\beta} = f_\alpha F_\alpha + f_\beta F_\beta = \frac{\left(\phi_\beta - \phi_0\right)F_\alpha + \left(\phi_0 - \phi_\alpha\right)F_\beta}{\phi_\beta - \phi_\alpha} \tag{4.21}$$

This expression has the form of a linear equation and gives a straight line when plotted with ϕ_0 not as a single composition point but as variable in the interval $[\phi_\alpha; \phi_\beta]$.

The energy of the mixed state, $F_{\text{mix}}(\phi_0)$, can be estimated based on the energy *change* upon mixing, as expressed by the Flory–Huggins formula from the last chapter (eq. 4.13). We substitute $\phi_A = \phi$ and $\phi_B = 1 - \phi$ and get

$$\Delta\bar{F}_{\text{mix}} = k_B T \left(\frac{\phi}{N_A} \ln \phi + \frac{(1-\phi)}{N_B} \ln(1-\phi) + \chi\phi(1-\phi) \right) \tag{4.22}$$

This expression has the form of a curve when plotted.[43]

Both functions, $F_{\text{mix}}(\phi)$[44] and $F_{\alpha\beta}(\phi)$, are shown in Figure 52 for the case of a system where demixing is more favorable (Panel A) and one where mixing is more favorable (Panel B). We can see for the phase-separated system in Panel (A) that the curve of the mixed state at composition ϕ_0, where we have a value of F_{mix}, lies above the straight line of the demixed state, where we have a value of $F_{\alpha\beta}$. Thus, when mixed initially, the system can spontaneously lower its energy by phase separation and thereby drop from F_{mix} to $F_{\alpha\beta}$ at the composition ϕ_0. We will then have two separate phases, α and β, with separate free energies F_α and F_β, which have

---

43 The energy change expressed by eq. (4.22) adds upon the initial state, which is the phase-separated one at an energy of $F_{\alpha\beta}(\phi_0)$, when mixing occurs, with either positive or negative sign depending on whether the mixing is disfavorable (positive sign) or favorable (negative sign). After the mixing, we therefore have a free energy of $F_{\text{mix}}(\phi) = F_{\alpha\beta}(\phi) + \Delta F_{\text{mix}}(\phi)$.

44 This is the function just "derived" in the latter footnote: $F_{\text{mix}}(\phi) = F_{\alpha\beta}(\phi) + \Delta F_{\text{mix}}(\phi)$. It consists of a summand $\Delta F_{\text{mix}}(\phi)$, which is the Flory–Huggins formula (4.22), which is added to (and therefore "sits upon") the straight line of the phase-separated state, $F_{\alpha\beta}(\phi)$ (4.21). At conditions for favorable mixing, the curve lays below the straight line as the $\Delta F_{\text{mix}}(\phi)$-summand has negative values, and vice versa.

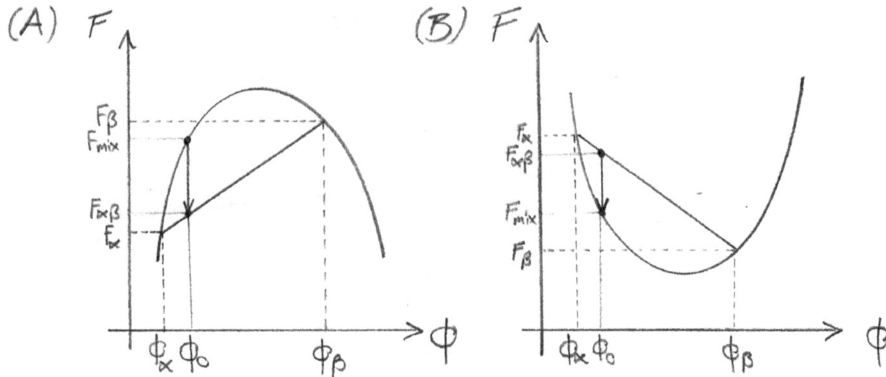

Figure 52: Free energy, *F*, as a function of the volume fraction of a component of interest, ϕ, in mixed or phase-separated polymer systems. In both plots, the straight line corresponds to the phase-separated state, whereas the curve corresponds to the mixed state (it represents the Flory–Huggins equation). In **(A)**, the concave curve lies above the straight line in a region between compositions ϕ_α and ϕ_β, meaning that the mixed state is less favorable than the phase-separated state there. As a result, a mixed system that has a composition ϕ_0 will spontaneously decompose into separated phases α and β, in which the component of interest is present in volume fractions ϕ_α and ϕ_β, respectively. In **(B)**, the convex curve lies below the straight line in a region between compositions ϕ_α and ϕ_β, meaning that the mixed state is more favorable that the phase-separated state. Pictures redrawn from M. Rubinstein, R. H. Colby: *Polymer Physics*, Oxford University Press, **2003**.

an average value of $F_{\alpha\beta}$. The straight line that denotes the demixed state in Figure 52 (A) connects these points, whose abscissa values denote the volume fractions of our component of interest A in the two phases, which are ϕ_α in the α-phase and ϕ_β in the β-phase. The relative fractions of the two phases are given by the length of the two parts of the straight line left and right of the point at $(\phi_0;F_{\alpha\beta})$, commonly referred to as the *lever rule*. In full contrast to all that, we can see for the mixed system in Panel (B) that the curve of the mixed state at composition ϕ_0, where we have an energy of F_{mix}, lies below the straight line of the demixed state, where we have an average energy of $F_{\alpha\beta}$ at the composition ϕ_0, which is actually a mean of two separated phases α and β with separate free energies F_α and F_β and volume fractions of our component A of ϕ_α in the α-phase and ϕ_β in the β-phase. In that situation, the system can spontaneously lower its energy by mixing and drop from the $F_{\alpha\beta}(\phi_0)$ line to the $F_{mix}(\phi_0)$ curve.

We can delimit the two different scenarios in Figure 52 by analyzing the curvature of the ΔF_{mix}-curve. This can be calculated by the second derivative of ΔF_{mix}. In case of the phase-separated system, the curve has a concave curvature. This means that the second derivative of ΔF_{mix} is negative, $\partial^2 \Delta F_{mix}/\partial\phi^2 < 0$. The mixed state, by contrast, exhibits a convex curvature, and the second derivative of ΔF_{mix} is positive, $\partial^2 \Delta F_{mix}/\partial\phi^2 > 0$.

Let us look at some boundary-case examples to understand this better. For an ideal mixture, the enthalpic contribution is zero, that is, $\Delta H_{\text{mix}} = 0$ or $\Delta U_{\text{mix}} = 0$. In that case, only the temperature-dependent entropic contribution remains (see eq. 4.1). The second derivative of ΔF_{mix} then is

$$\frac{\partial^2 \Delta \bar{F}_{\text{mix}}}{\partial \phi^2} = -T\frac{\partial^2 \Delta \bar{S}_{\text{mix}}}{\partial \phi^2} = k_{\text{B}}T\left(\frac{1}{N_{\text{A}}\phi} + \frac{1}{N_{\text{B}}(1-\phi)}\right) > 0 \tag{4.23}$$

The outcome is a positive value at any ϕ, which means that $\Delta F_{\text{mix}}(\phi)$ has a convex curvature. Hence, mixing is always favorable, which is reasonable: entropy always favors mixing independent of the mixture's composition.

In an opposite example, let us ignore the entropy term and only look at the energetic contributions. The second derivative of ΔF_{mix} then is

$$\frac{\partial^2 \Delta \bar{F}_{\text{mix}}}{\partial \phi^2} = \frac{\partial^2 \Delta \bar{U}_{\text{mix}}}{\partial \phi^2} = -2\chi k_{\text{B}}T \tag{4.24}$$

This represents the findings that we have derived in the last chapter. When $\chi < 0$, which according to our discussion in Section 4.1.2 denotes that mixing is favored, then the curvature is convex, which also denotes mixing to be favored according to what we have just said above. When $\chi > 0$, which according to our discussion in Section 4.1.2 denotes that mixing is disfavored, then the curvature is concave, which also denotes mixing to be disfavored according to what we have just said above.[45]

In an actual mixture, both the energetic and the entropic term are of course relevant:

$$\frac{\partial^2 \Delta \bar{F}_{\text{mix}}}{\partial \phi^2} = \frac{\partial^2 \Delta \bar{U}_{\text{mix}}}{\partial \phi^2} - T\frac{\partial^2 \Delta \bar{S}_{\text{mix}}}{\partial \phi^2} = k_{\text{B}}T\left(\frac{1}{N_{\text{A}}\phi} + \frac{1}{N_{\text{B}}(1-\phi)}\right) - 2\chi k_{\text{B}}T \tag{4.25}$$

Let us look at how the curves in an $F(\phi)$ diagram look like in that situation.

45 We may get a further insight when we write out the temperature-dependent form of χ, namely $\chi = \chi_S + \chi_H/T$, in eq. (4.24):

$$\frac{\partial^2 \Delta \bar{F}_{\text{mix}}}{\partial \phi^2} = \frac{\partial^2 \Delta \bar{U}_{\text{mix}}}{\partial \phi^2} = -2\chi k_{\text{B}}T = -2\left(\chi_S + \frac{\chi_H}{T}\right)k_{\text{B}}T = -2k_{\text{B}}T\chi_S - 2k_{\text{B}}\chi_H$$

For $T \to 0$, this leads to $\frac{\partial^2 \Delta \bar{F}_{\text{mix}}}{\partial \phi^2} = \frac{\partial^2 \Delta \bar{U}_{\text{mix}}}{\partial \phi^2} = -2k_{\text{B}}\chi_H \stackrel{\text{def}}{=} -z \cdot (2u_{\text{AB}} - u_{\text{AA}} - u_{\text{BB}})$

From this, we see that

- if $u_{\text{AB}} > \frac{u_{\text{AA}} + u_{\text{BB}}}{2}$, which corresponds to an unstable mixture, then we have $\chi_H > 0$
- if $u_{\text{AB}} < \frac{u_{\text{AA}} + u_{\text{BB}}}{2}$, which corresponds to a stable mixture, then we have $\chi_H < 0$

Figure 53 displays an example of an unsymmetric polymer blend ($N_A \neq N_B$). Shown is the change of the free energy of mixing, ΔF_{mix}, as a function of the volume fraction of component A, ϕ, at different temperatures. At high temperature, the entropy term dominates and creates a global minimum. The curve is convex allover, which means that mixing is favored at all compositions. At low temperature, by contrast, the energy term that usually disfavors mixing creates a miscibility gap. In this region, the curve is *locally* concave, and the system can decrease F by phase separation down to a value represented by the straight line that connects the energy minima. This line is the **common tangent**, because that kind of line is the deepest possible one touching the curve such that $\partial F/\partial\phi$ is the same for the curve and the straight line in the two phase-coexistence points, respectively. This must be the case, as $\partial F/\partial\phi$ corresponds to the chemical potential, which must be the same in the coexistence points ϕ' and ϕ'' to have equilibrium there.

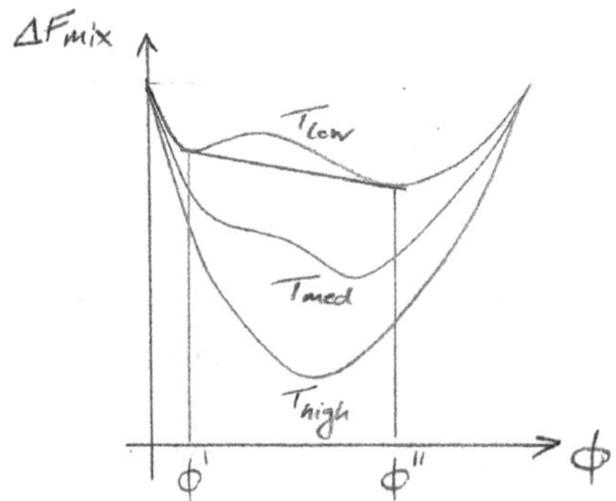

Figure 53: Free energy of mixing, ΔF_{mix}, as a function of the volume fraction of a component of interest, ϕ, in a polymer system at different temperatures. At high temperature, dominance of the entropy term creates a distinct minimum, thereby favoring mixing at all compositions. At low temperature, the energy term that usually disfavors mixing creates a miscibility gap at intermediate compositions.

According to this latter thought, the common tangent delimits the stable "rim-ϕ regions" from the unstable "mid-ϕ region". However, a "bad" concave curvature is present actually only in the innermost-ϕ region, between the inflection points. Hence, we can delimit three different regimes. First, we have the rim-ϕ regime beyond the common-tangent touching points. This corresponds to two regions, one at the left ϕ-rim and the other at the right ϕ-rim, that both enable **stable** mixtures. In contrast, the second, innermost-ϕ regime between the inflection points delimits the region in which the mixture is truly **unstable**. The third domain lies between these two extremes, and is called the **metastable** domain. Here, we have a convex curvature, indicating favorable mixing, but we are already above the deepest possible straight line, the common tangent, which indicates unfavorable mixing.

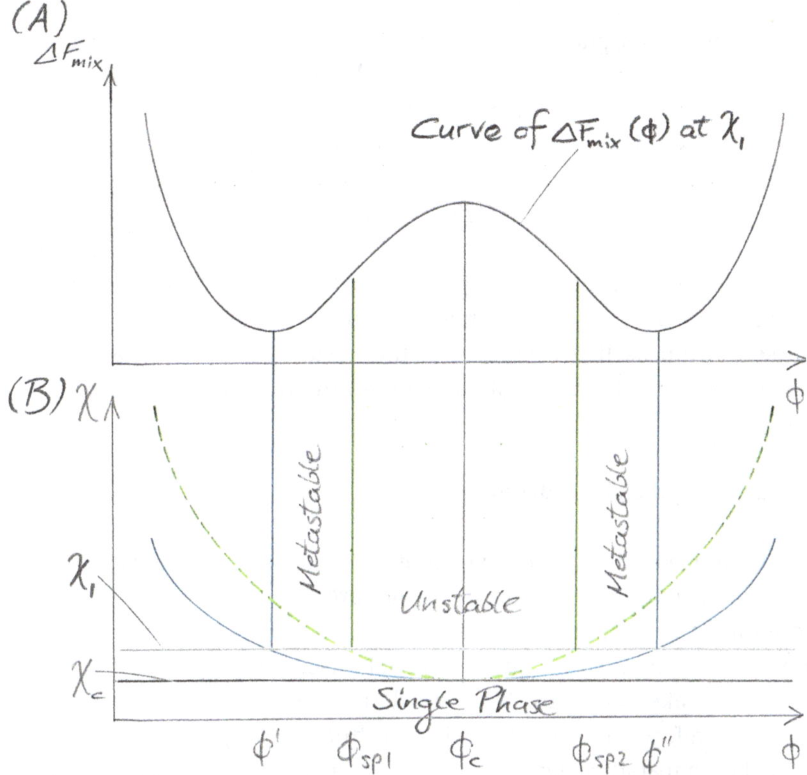

Figure 54: The miscibility gap is further subdivided into unstable and metastable domains. (**A**) At the edges of the miscibility-gap region, the curvature of the $F(\phi)$-curve is convex, which would correspond to a stable mixture, but we still have $F_{mix} > F_{phase\text{-}sep.}$ there. As a result, this is a metastable domain. (**B**) The separation between the single phase and the metastable regime is called the *binodal line*, whereas the separation between the metastable and the unstable regime is called the *spinodal line*. Both these lines coincide at the *critical point* $(\chi_c;\phi_c)$.

We can take a closer look at the miscibility gap by considering a symmetric polymer blend with two components that have the same N, as shown in Figure 54(A). Between the inflection points, that is, for the volume fractions ϕ_{sp1}–ϕ_{sp2}, the curvature is concave $\left(\partial^2 \Delta F_{mix}/\partial \phi^2 < 0\right)$. The mixture is unstable, and even smallest fluctuations induce phase separation, which is called **spinodal decomposition.** Between the inflection points and the common-tangent touching points,[46] for the volume fractions ϕ'–ϕ_{sp1} and ϕ_{sp2}–ϕ'', the curvature is convex $\left(\partial^2 \Delta F_{mix}/\partial \phi^2 > 0\right)$, but still $F_{mix}(\phi_0) > F_{\alpha\beta}(\phi_0)$. In this regime, the mixture is metastable, which means that it is

46 In the case of a symmetric blend, the common tangent is a horizontal line with slope zero, so its touching points with the $\Delta F_{mix}(\phi)$-curve are the minima of that curve.

locally stable against small fluctuations, whereas phase separation will set in at larger fluctuations. In other words, phase separation needs **nucleation and growth** here.

Figure 54(B) shows the same system in a plot of the Flory–Huggins parameter, χ, as a function of the polymer volume fraction, ϕ. This is a phase diagram. At low χ, the components like or at least tolerate each other, and mixing is possible at all compositions. To the contrary, at high χ, the components do not like each other, such that demixing will occur at certain compositions, delimited by the same boundaries as just discussed for the $F(\phi)$-plot above. The point where this can first happen is called the **critical point** at $\chi_c;\phi_c$.[47] The regions of nonstability are delimited by two lines. The **binodal line** separates the stable from the nonstable domain. The nonstable domain itself is composed of the unstable and the metastable regime, which are delimited from each other by the **spinodal line**.

4.2.2 Construction of the phase diagram

So far, we have derived a diagram of F versus ϕ with a set of T-dependent curves to estimate at what composition, at a given temperature, the system is mixed or demixed. More intuitive, though, would be a diagram with two practical variables, T and ϕ, in which we can directly map domains of miscibility and immiscibility. To construct this, we take two steps. First, we derive a χ–ϕ diagram as plotted in Figure 54(B). Second, we derive a T–ϕ diagram from the χ–ϕ diagram by taking into account the temperature dependence of the Flory–Huggins parameter χ. We begin by calculating the binodal and spinodal lines of the first diagram.

As a toolkit, we need the energy of mixing given by

$$\Delta\bar{F}_{\text{mix}} = k_B T\left(\frac{\phi}{N_A}\ln\phi + \frac{(1-\phi)}{N_B}\ln(1-\phi) + \chi\phi(1-\phi)\right) \tag{4.26a}$$

its first derivative given by

$$\frac{\partial\Delta\bar{F}_{\text{mix}}}{\partial\phi} = k_B T\left(\frac{\ln\phi}{N_A} + \frac{1}{N_A} + \frac{\ln(1-\phi)}{N_B} + \frac{1}{N_B} + \chi(1-2\phi)\right) \tag{4.26b}$$

and its second derivative given by

$$\frac{\partial^2\Delta\bar{F}_{\text{mix}}}{\partial\phi^2} = k_B T\left(\frac{1}{N_A\phi} + \frac{1}{N_B(1-\phi)} - 2\chi\right) \tag{4.26c}$$

47 This point is conceptually identical to the critical point in a one-component phase diagram and the related van der Waals equation at $(p_c;T_c;V_c)$, at which liquefaction of a gas can first occur.

The phase boundary, or the binodal line, is given by the common tangent. For simplicity, we consider a symmetrical blend with $N_A = N_B = N$. In this case, the common tangent is a horizontal line with slope zero. This can be calculated by setting the first derivative of F_{mix} to zero:

$$\left(\frac{\partial \Delta \bar{F}_{\text{mix}}}{\partial \phi}\right)_{\substack{\phi=\phi' \\ \phi=\phi''}} = k_B T\left(\frac{\ln \phi}{N} + \frac{\ln(1-\phi)}{N} + \chi(1-2\phi)\right) \stackrel{!}{=} 0 \tag{4.27}$$

Rearrangement for χ yields the *binodal* $\chi(\phi)$ curve:

$$\chi_{\text{Binodal}} = \frac{1}{2\phi - 1}\left(\frac{\ln \phi}{N} - \frac{\ln(1-\phi)}{N}\right) = \frac{\ln\left(\frac{\phi}{1-\phi}\right)}{(2\phi-1)N} \tag{4.28}$$

The inflection points delimit the metastable from the unstable regime. They can be calculated by setting the second derivative of F_{mix} to zero:

$$\frac{\partial^2 \Delta \bar{F}_{\text{mix}}}{\partial \phi^2} = k_B T\left(\frac{1}{N\phi} + \frac{1}{N(1-\phi)} - 2\chi\right) \stackrel{!}{=} 0 \tag{4.29}$$

Rearrangement for χ yields the *spinodal* χ (ϕ) curve:

$$\chi_{\text{Spinodal}} = \frac{1}{2}\left(\frac{1}{N\phi} + \frac{1}{N(1-\phi)}\right) \tag{4.30}$$

The minimum of both the latter curves denotes the *critical point*, $(\chi_c;\phi_c)$. This point marks the very first possibility for demixing. Below χ_c, mixing is possible for all volume fractions. We can calculate the critical point by setting the first derivative of the latter equation for the spinodal line to zero to estimate its minimum:

$$\frac{\partial \chi_{\text{Spinodal}}}{\partial \phi} = \frac{1}{2}\left(\frac{1}{N\phi^2} + \frac{1}{N(1-\phi)^2}\right) \stackrel{!}{=} 0 \tag{4.31a}$$

At this point of our discussion, we can actually drop our above simplification and consider the general case of a nonsymmetric blend again. This is mathematically imprecise, but rather based on a logical argument. Nevertheless, let us just go this way: the N in the first summand belongs to the volume fraction ϕ of compound A (note that in this section, we write ϕ for what we have earlier denoted as ϕ_A), whereas the N in the second summand refers to the volume fraction $1 - \phi$ (i.e., $1 - \phi$), which is that of compound B (because $1 - \phi_A = \phi_B$). So, we get

$$\frac{\partial \chi_{\text{Spinodal}}}{\partial \phi} = \frac{1}{2}\left(\frac{1}{N_A\phi^2} + \frac{1}{N_B(1-\phi)^2}\right) \stackrel{!}{=} 0 \tag{4.31b}$$

Solving this equation for ϕ yields the **critical composition,** ϕ_c:

$$\phi_c = \frac{\sqrt{N_B}}{\sqrt{N_A} + \sqrt{N_B}} \tag{4.32}$$

As we can see, a symmetric blend with $N_A = N_B$ has a critical composition of $\phi_c = 0.5$. This means that at this composition, demixing can first occur if χ is too unfavorable and drops below χ_c. This critical χ_c can be quantified by inserting ϕ_c for ϕ in eq. (4.30):

$$\chi_c = \frac{1}{2}\left(\frac{1}{\sqrt{N_A}} + \frac{1}{\sqrt{N_B}}\right)^2 \tag{4.33}$$

Let us examine the latter equation in some detail.

In a regular small-molecule solution, where $N_A = N_B = 1$, the **critical interaction parameter** is $\chi_c = 2$. As a result, everything with $\chi < \chi_c = 2$ will mix, whereas everything with $\chi > \chi_c = 2$ may demix if ϕ is around ϕ_c.

In a polymer solution, with $N_A >> 1$ and $N_B = 1$, the picture is different. Here, $\chi_c = ½$, which is much lower than in the small-molecule scenario. Here, everything with $\chi < \chi_c = ½$ will mix, whereas everything with $\chi > \chi_c = ½$ may demix if ϕ is around ϕ_c. A polymer solution in which χ is exactly ½ is in the Θ-state, which means at the borderline between miscibility and immiscibility.

In a polymer blend, where both $N_A >> 1$ and $N_B >> 1$, the critical Flory–Huggins parameter is basically zero, $\chi_c \approx 0$. As a result, mixing is hardly possible at all in this case, as for mixing to occur, χ must be smaller than χ_c, which is hard to achieve when $\chi_c \approx 0$ already. Mixing two polymers therefore practically only works for $\chi = 0$, which is very rare! We therefore have to resort to other strategies than mixing if the properties of different polymers shall be combined. A common way to achieve that is by copolymerization of their different monomers in one copolymer species ("mixing within the chain"), or by incorporation of attractive side groups in the to-be-mixed polymers that aid to overcome the enthalpy penalty of mixing.

As a numerical example, we calculate the volume fractions at the phase boundaries of the spinodal curve of a polymer solution at a degree of polymerization of $N_A = 1000$ with a Flory–Huggins interaction parameter of $\chi = 1.5$; hence, we consider a system beyond χ_c that already exhibits a decent two-phase regime. The spinodal curve separates the two-phase regime from the metastable coexisting regime. Mathematically, it is determined by the set of inflection points of the function $\Delta F_{mix}(\phi_A)$; these points can be calculated by setting the second derivative of this function to zero. The resulting polymer volume fractions then correspond to the polymer concentrations of the two separate phases that arise from the phase separation.[48] For our example, the calculation goes as follows:

$$\frac{\Delta F_{\text{mix}}}{k_{\text{B}}T} = \frac{\phi_{\text{A}}}{N_{\text{A}}} \cdot \ln \phi_{\text{A}} + \phi_{\text{B}} \cdot \ln \phi_{\text{B}} + \chi \phi_{\text{A}} \phi_{\text{B}} = \frac{\phi_{\text{A}}}{N_A} \cdot \ln \phi_{\text{A}} + (1-\phi_{\text{A}}) \cdot \ln(1-\phi_{\text{A}}) + \chi \phi_{\text{A}}(1-\phi_{\text{A}})$$

The first derivate of that function is:

$$\frac{\partial \left(\frac{\Delta F_{\text{mix}}}{k_{\text{B}}T}\right)}{\partial \phi_{\text{A}}} = \frac{1}{N_{\text{A}}} \cdot \frac{\phi_{\text{A}}}{\phi_{\text{A}}} + \frac{1}{N_{\text{A}}} \cdot \ln \phi_{\text{A}} + (1-\phi_{\text{A}}) \cdot \frac{-1}{(1-\phi_{\text{A}})} - \ln(1-\phi_{\text{A}}) + \chi - 2\chi\phi_{\text{A}}$$

$$= \frac{1}{N_{\text{A}}} + \frac{1}{N_{\text{A}}} \cdot \ln \phi_{\text{A}} - 1 - \ln(1-\phi_{\text{A}}) + \chi - 2\chi\phi_{\text{A}}$$

We further calculate the second derivative, plug in the numbers from the example and set the entire equation equal to zero:

$$\frac{\partial^2 \left(\frac{\Delta F_{\text{mix}}}{k_{\text{B}}T}\right)}{\partial {\phi_{\text{A}}}^2} = \frac{1}{N_{\text{A}}} \cdot \frac{1}{\phi_{\text{A}}} - \frac{-1}{(1-\phi_{\text{A}})} - 2\chi = \frac{1}{1000} \cdot \frac{1}{\phi_{\text{A}}} + \frac{1}{(1-\phi_{\text{A}})} - 2 \cdot 1.5 \overset{!}{=} 0$$

Further rearrangement yields:

$$\frac{1}{1000} \cdot (1-\phi_{\text{A}}) + \phi_{\text{A}} - 3(1-\phi_{\text{A}})\phi_{\text{A}} = 0 \Leftrightarrow 3{\phi_{\text{A}}}^2 + \left(1 - \frac{1}{1000} - 3\right)\phi_{\text{A}} + \frac{1}{1000} = 0$$

$$\Leftrightarrow {\phi_{\text{A}}}^2 - 0.667\phi_{\text{A}} + 0.0003333 = 0$$

We have now generated a quadratic equation in its reduced form and can solve for its two solutions using the *p*–*q* formula:

$$\phi_{\text{A}} = \frac{0.667}{2} \pm \sqrt{\left(\frac{0.667}{2}\right)^2 - 0.0003333} = \frac{0.667}{2} \pm 0.333 = 0.6665 \text{ and } 0.0005$$

As we can see from the resulting volume fractions, spinodal decomposition does in fact not produce pure solvent and pure polymer phases, but instead, a polymer-rich phase (here: 67% polymer) and a polymer-lean phase (here: 0.5‰ polymer) are obtained. This means that the phase containing most of the polymer still contains a lot of solvent (here: 33%), and thus needs to be properly dried to obtain the pure polymer.

We have learned in Section 4.1.3 that the Flory–Huggins parameter is not a constant, but rather a temperature-dependent function according to eq. (4.14a). This allows us to translate the χ–ϕ representations we have discussed so far (Figure 54(B)) into more practical T–ϕ diagrams, as shown in Figure 55. These can have two very different

48 Note that these volume fractions are temperature dependent due to the temperature dependence of the Flory–Huggins parameter $\chi(T)$, as will be discussed in the next paragraph.

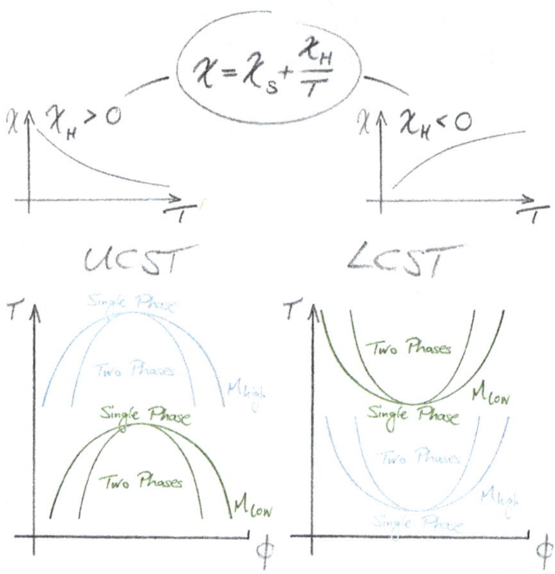

Figure 55: The sign of the enthalpy-component of the *T*-dependent Flory–Huggins parameter determines whether mixing is favorable or disfavorable at high or low *T*, which are two opposite scenarios assessed by either an upper critical solution temperature (UCST), as shown on the left, or a lower critical solution temperature (LCST), as shown on the right. In both schematics, we see that there is also an impact of the molar mass of the chains.

appearances, depending on the sign of the χ_H-contributor, corresponding to two types of miscibility. In the first case, depicted on the left-hand side of Figure 55, $\chi_H > 0$, which means that the overall χ parameter decreases as the temperature rises. These systems display a volume-fraction-dependent **upper critical solution temperature (UCST)**, above which mixing is possible. In the second case, $\chi_H < 0$, and the overall χ parameter increases with rising *T*. Such mixtures exhibit a volume-fraction-dependent **lower critical solution temperature (LCST)**, above which mixing is *im*possible!

A UCST is the usual case for many polymer–solvent mixtures without specific intercomponent interactions such as mutual hydrogen bonding, as we have discussed in Section 4.1.3. The exact value of the UCST is also molar-mass dependent, as can also be seen in Figure 55. In the UCST graph, we see that upon decrease of temperature, an accompanying decrease of the solvent quality such that χ exceeds ½, and with that, an onset of precipitation is encountered for long chains sooner than for short chains. Analogously, in the LCST graph, we see that upon increase of temperature, an accompanying decrease of the solvent quality such that χ exceeds ½, and with that, an onset of precipitation is encountered for long chains sooner than for short chains. In both cases, this circumstance can be used for **fractionation of polydisperse samples**, simply by dissolving such a sample in a solvent and then gradually worsening the solvent quality, either by suitable change of temperature or by successive addition of a nonsolvent. Upon that, first the longest chains precipitate, then the next shorter fraction, and so forth. Isolating these fractions before precipitating the next thereby allows the broadly polydisperse sample to be fractionated into several less polydisperse ones. This may either be used preparatively or at least analytically to determine a coarse-grained molar-mass distribution by analyzing the average molar mass and the amount of each fraction through means like light scattering, osmometry, and gravimetry.

More complex phase diagrams may arise in special cases due to the sometimes complex temperature dependence of the Flory–Huggins interaction parameter that might change its sign multiple times. Figure 56 shows some of these special cases. Panel (A) is for a mixture with *both* a UCST *and* an LCST, Panel (B) is for a mixture with a closed miscibility gap, and Panel (C) displays an hourglass-shaped phase diagram. The first case has a rather simple explanation, shown in a χ–T representation in Panel (D). Initially, we see a regular Flory–Huggins-type UCST upon increase of temperature. On top of that, however, comes another effect: an increase of temperature often leads to thermal expansion of both the solvent and the polymer. This expansion is usually more pronounced for the solvent than for the polymer. Hence, in our lattice picture, we have to increase the lattice-site volume for the solvent more than for the polymer's repeating units upon increase of temperature, meaning that we need to distort our lattice to fit them both on, which entails an entropy penalty. If this penalty is too marked, the system will phase-separate again, creating an LCST in the process. Hence, we actually always have both an LCST and a UCST, and depending on where they and their two binodal curves lie, we may

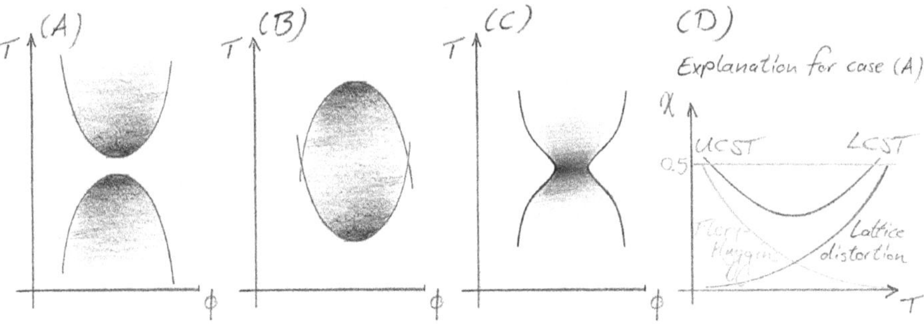

Figure 56: Complex shapes of phase diagrams that result from the interplay of the T-dependent χ-parameter in the enthalpy term and the T-dependent $T\Delta S$ term in the entropy term of the Flory–Huggins equation. Case **(A)** has an illustrative explanation, as visualized in **(D)**: the usual UCST-type T-dependence of χ leads to a decrease of it upon raise of T (making mixing more favorable, or better saying "less disfavorable" at high T), whereas the different thermal expansion of the solvent and the polymer that comes along with a temperature increase causes energy and entropy penalties that oppose this; together, this creates a minimum of $\chi(T)$, corresponding to a closed miscibility gap in the phase diagram. The sketch in Panel **(D)** is inspired by J.M.G. Cowie: *Chemie und Physik der synthetischen Polymeren*, Vieweg, 1991.

have phase diagrams of kind (A), (B), or (C). The reason why we often nevertheless (seemingly) observe simple phase diagrams such as those shown in Figure 55, with just one critical temperature (LCST *or* UCST, but not both of them) is that many polymer–solvents systems actually exhibit both, but one of them is usually "hidden", either in the form of a UCST that lies below the melting point of the solvent, or in the form of an LCST that is located above the boiling point of the solvent.

4.2.3 Mechanisms of phase transitions

4.2.3.1 Spinodal decomposition

In the unstable two-phase regime (see Figure 54(B)), any random concentration fluctuation self-amplifies and eventually leads to macroscopic phase separation. This is because the gradient of the chemical potential μ is opposite to the gradient of concentration in this regime. In very general, concentration fluctuations in a system are equilibrated by a diffusive flux of matter along the gradient of the chemical potential μ, which is defined as $\mu = (\partial G/\partial n) = (\partial G/\partial \phi)$ and exhibits a monotonic concentration dependence in mixed phases: $\mu_{mix} = \mu_0 + RT \ln \phi$. The volume fraction ϕ in the second equation necessitates a high chemical potential at high concentrations, and conversely, a low chemical potential at low concentrations. If there is a gradient in the chemical potential, $(\partial \mu/\partial \phi) = (\partial^2 G/\partial \phi^2) > 0$, diffusion leads to an overall flux of matter from the higher to the lower potential until the gradient is equilibrated. Normally, such a chemical-potential gradient is in line with the concentration gradient, so the diffusive flux goes from high to low concentrations, referred to as **downhill diffusion**. In the two-phase spinodal decomposition regime, by contrast, the gradient of the chemical potential is *reverse* to that in concentration, $(\partial \mu/\partial \phi) = (\partial^2 G/\partial \phi^2) < 0$. Diffusion still takes place along the chemical-potential gradient, but now this means it occurs *against* the concentration gradient. The resulting self-amplification of concentration differences is called **uphill diffusion** and eventually leads to a macroscopic phase separation, the so-called **spinodal decomposition**. This phenomenon occurs on the characteristic lengthscale of the concentration fluctuations in the system. If that lengthscale is large, diffusion has to cover large distances, which is unlikely; if the lengthscale is short, many new polymer–solvent interfaces must be created, which is disfavored, too. Therefore, spinodal decomposition occurs on an intermediate lengthscale that constitutes the best compromise between the two effects. This lengthscale remains constant at first and is amplified only in its amplitude. During later stages of the process, the lengthscale diverges, and the phase separation actually becomes macroscopic. This last phase can be observed, for example, by optical microscopy, if one of the components is colored. Earlier stages can be observed by complementary methods such as neutron or X-ray scattering.

4.2.3.2 Nucleation and growth

In the metastable regime (again, see Figure 54(B)), the system is able to tolerate small concentration fluctuations, and thus, spinodal decomposition is prohibited. Here, a different phase-separation mechanism is in action: **nucleation and growth**. Instead of a macroscopic phase separation, small polymer-rich phase clusters form locally. As can be seen in Figure 57, these clusters need to grow to a specific minimal size with a radius r^* for the nucleation process to be energetically favorable. This is because the unfavorable interfacial energy of the nuclei first is stronger than their favorable volume energy, but this ratio reverses from a cluster minimal size r^* on. If the nuclei do not make it to that size, nucleation does not surpass the minimum size for macroscopic phase separation, so the system can remain two-phased. By contrast, once the minimal size r^* is reached, macroscopic phase separation sets in. Therefore, impurities with a radius bigger than r^*, such as dust particles, often function as a seed for nucleation. Alternatively, artificial nucleation seeds can be added on purpose to facilitate the process. This is called *hetero-nucleation*. Once the critical nucleation size is exceeded and a nucleus is formed, more material adds to it easily, and the new phase grows.

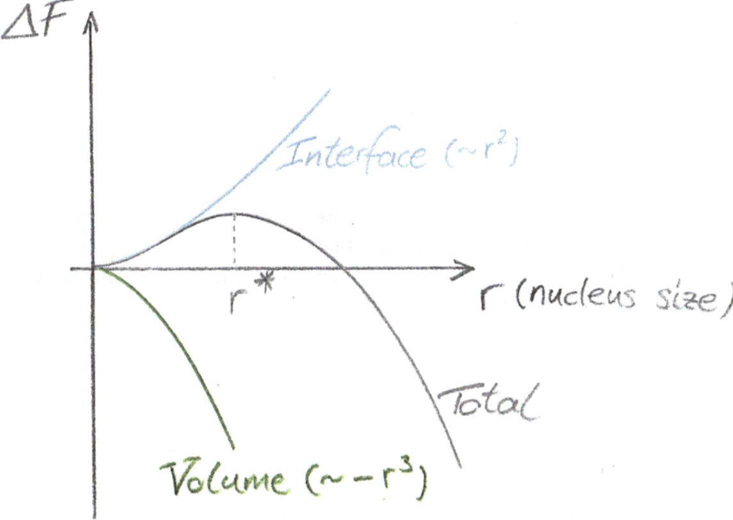

Figure 57: Graphical representation of the free energy of nucleation. Two factors contribute to the free energy: the interface and the volume of the nucleus. Nucleation is energetically favorable only after the cluster has grown to a minimal size r^*.

Figure 58 summarizes the two polymer phase-separation mechanisms in view of the system's state of stability or instability. In the unstable regime, spinodal decomposition is the mode of phase separation. In the metastable regimes, phase separation happens through nucleation and growth.

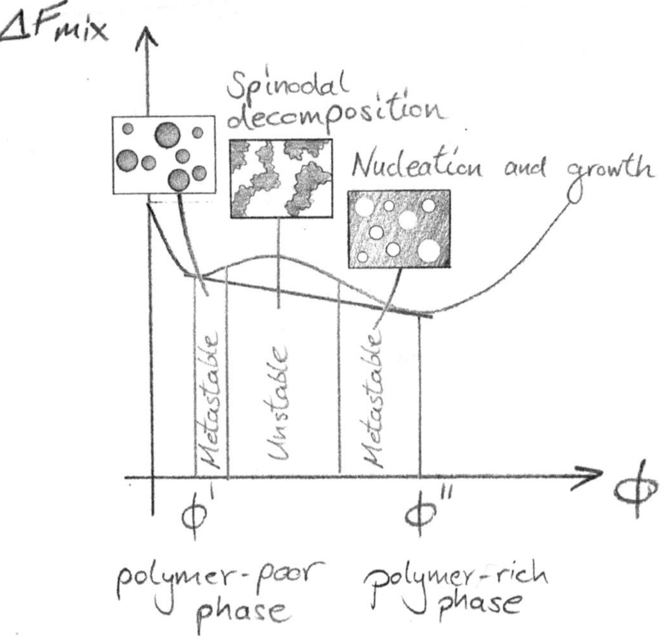

Figure 58: Schematic representation of phase separation regimes of polymer–solvent or polymer–polymer mixtures. Shown is the mixing free energy as a function of the polymer volume fraction, like in Figure 53. In the unstable regime, spinodal decomposition takes place. In the metastable regimes, phase separation occurs by nucleation and growth.

4.3 Osmotic pressure

LESSON 9: OSMOTIC PRESSURE

You have got to know the Flory–Huggins parameter as an elementary tool to appraise whether or not polymers are miscible with solvents or other polymers. What you are lacking, though, is an experimental technique to determine this parameter. Such a technique will be introduced in the following lesson. You will see that in addition to quantifying interactions between a polymer and its surrounding, and therewith determining the Flory–Huggins parameter, this method also allows you to determine the molar mass of a polymer. Furthermore, you will again see the striking similarity of the physical-chemical concepts behind real polymer chains and real gases.

So far we have looked at polymer thermodynamics from a theoretical viewpoint and introduced the Flory–Huggins parameter χ as a simple means to quantify the polymer–solvent interaction and to predict the state of solvency of one in the other (meaning if it is a good-solution state, a θ-state, or a bad- or nonsolvent state). Earlier in this book, in Chapter 3, we have introduced a quantity that expresses pretty much the same: the excluded volume, v_e. Now, we will take a look on how to *determine* χ and v_e experimentally; this discussion will also show us how these two parameters are related to each other.

Consider the experimental setup that is shown in Figure 59. This setup features a vessel containing a polymer solution immersed in a bath with excess solvent. Both are separated by a semipermeable membrane that allows for solvent interdiffusion but blocks polymer interdiffusion, simply because its pores are too tight for the large polymer coils to diffuse through, whereas the small solvent molecules can. As a result, since the polymer cannot diffuse out of its vessel to spread over the whole system (and thereby equilibrate its chemical-potential gradient), in turn, part of the solvent will diffuse into the polymer vessel such to dilute the polymer solution in there. This influx of solvent into the polymer chamber creates a mechanical pressure, in the present case by raise of a fluid column in a vertical tube. Once this mechanical pressure, p, is equal to the osmotic pressure of the solution, Π, equilibrium is reached. At this state, the elevation of the fluid column in the vertical tube, Δh, has come to a final value. We may then write:

$$\Pi = p = \frac{m \cdot g}{\text{Area}} = \frac{V\rho \cdot g}{\text{Area}} = \frac{\pi r^2 \Delta h \rho \cdot g}{\pi r^2} = \Delta h \rho \cdot g \tag{4.34}$$

With eq. (4.34), we have a simple means to measure Π, simply by measuring Δh. Π, in turn, is a central parameter in the equation of state of an osmotic system, which is Van't Hoff's law for dilute solutions:

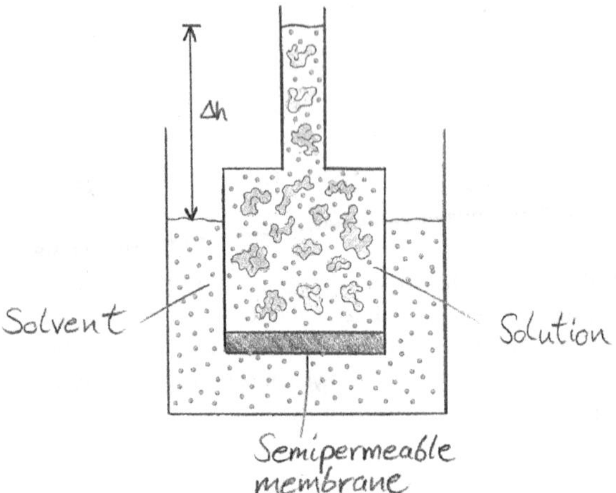

Figure 59: Experimental setup to determine the osmotic pressure of a polymer solution. The polymer cannot pass through the pores of a membrane separating it from excess surrounding solvent, whereas the solvent molecules can. As a result, influx of solvent into the solution-containing compartment will cause a mechanical pressure, for example, by rise of a fluid column in a tube. Equilibrium is reached when this mechanical pressure matches and balances the osmotic pressure of the solution. Picture inspired by B. Tieke: *Makromolekulare Chemie*, Wiley VCH, **1997**.

$$\Pi \cdot V = n \cdot RT \tag{4.35}$$

Rearrangement yields a relation between the osmotic pressure, Π, and the polymer mass-per-volume concentration, c:

$$\Pi = \frac{n \cdot RT}{V} = \frac{m}{M_n} \cdot \frac{RT}{V} = \frac{c \cdot RT}{M_n} \Rightarrow M_n = \frac{RT}{\Pi/c} \tag{4.36}$$

With that, we can determine the number-average molar mass of our polymer simply by a concentration-dependent osmometric experimental series.[49] Note that in the

49 In general, the number-average molar mass of a dissolved species can be determined by solution-based effects whose magnitude solely depends on the *number* of dissolved molecules. Such effects are known as *colligative properties* (derived from Latin collegere = to collect), the prime examples of which being the osmotic pressure, the boiling-point elevation, and the freezing-point depression of a solution. All these effects have a magnitude that is proportional to the number of dissolved molecules, and therefore they allow the molar amount of the dissolved compound in the solution to be determined. If the solution has been made from a known mass of that dissolved compound, then dividing the one by the other gives the weight per mole, and therewith, the molar mass. We may express this general principle quantitatively as follows. Let Y be a colligative property (i.e., one of the above three effects), whose extent is proportional to the number of dissolved molecules of species i per volume, N_i/V, with a constant of proportionality K: $Y = K\frac{N_i}{V}$. If there is several different species in the solution, for example, several different molar masses of a polymer

upper notation, the concentration $c = m/V$ has the unit $g \cdot L^{-1}$, as it is common in polymer science. In classical physical chemistry, by contrast, concentration is commonly defined as n/V with the unit $mol \cdot L^{-1}$. In this form, Van't Hoff's law would read $\Pi = c\,RT$ instead of the polymer-science variant noted in eq. (4.36).

Van't Hoff's law is valid for dilute solutions. Nondilute solutions deviate from that simple relation and are best described by a **virial series expansion** of it

$$\Pi/c = RT\left(\frac{1}{M_n} + A_2 c + A_3 c^2 + \ldots\right) \tag{4.37}$$

In the notation of eq. (4.37), the osmotic pressure is given in a reduced form, Π/c.[50] A key parameter in the above equation is the **second virial coefficient**, A_2. It accounts for deviations from the ideal Van't Hoff law due to interactions at higher concentration. Thus, A_2 is an important parameter to *quantify* such interactions. Often, experimental data can be well fitted to eq. (4.37) in a linear form, where the series is expanded only up to the linear A_2 term. Figure 60 sketches such a dataset and linear fit. From the intercept, RT/M_n, we get the polymer's number-average molar mass, and from the slope, $RT \cdot A_2$, we get the second virial coefficient A_2.

The second virial coefficient quantifies the interaction between a polymer and a solvent. With that, it is closely connected to the Flory–Huggins parameter χ, as we will discuss in more detail below. Generally, a positive A_2-value means good solvation of the polymer, whereas a negative A_2-value means nonsolubility; in the Θ-state,

---

with a polydisperse distribution, then we get $Y = K\frac{\sum_i N_i}{V}$. In polymer science, a typical measure of concentration is mass per volume, that is, $c = \frac{\sum_i w_i}{V} = \frac{\sum_i N_i M_i}{N_A V}$. If we normalize our colligative property Y by the concentration c (see the next footnote for some background on this way of normalization), we obtain: $\frac{Y}{c} = K\frac{\sum_i N_i}{V} \cdot \frac{N_A V}{\sum_i N_i M_i}$. Together with the definition of the number-average molar mass, $M_n = \frac{\sum_i N_i M_i}{\sum_i N}$, this yields: $\frac{Y}{c} = \frac{K N_A}{M_n}$. We see from that identity that all colligative properties are connected to the number-average molar mass M_n, and therefore they all allow it to be determined experimentally. The extent of how strongly a colligative property Y depends on N is set by the constant of proportionality K. For the osmotic pressure, this constant is particularly high, so the effect is very strong for this kind of colligative property. This is the reason why it is favored as a method in polymer science: in a given mass-amount of polymer sample we only have a few molecules (but each one very large in turn), and so we need a colligative property where even such few (large) molecules have a strong effect that is well measurable. Osmotic pressure is such a favorable property. On top of that, there is another reason why this method is beloved by polymer scientists: in addition to M_n, it also gives quantitative information on the quality of polymer–solvent interactions. This is what the following discussion in this section is about.

50 A reduced quantity is one that is normalized by concentration. Other such normalized quantities are known to you from elementary physical chemistry already, for example, specific quantities (normalized by mass) or molar quantities (normalized by the molar number).

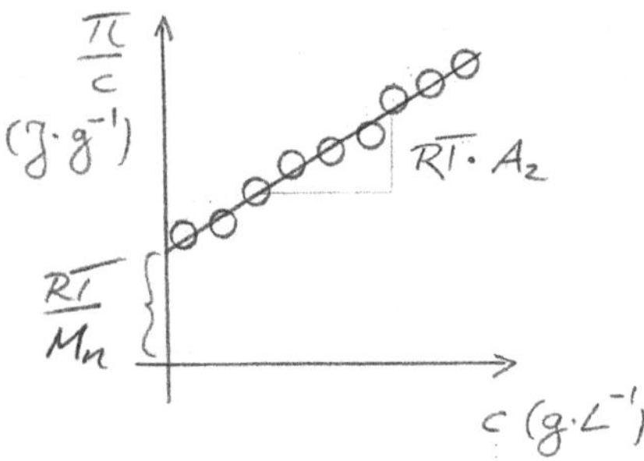

Figure 60: Plot of the series expansion of the concentration-dependent osmotic pressure in its typical form.

A_2 is zero.[51] Figure 61 schematizes typical data for poly(methyl methacrylate) in various solvents as an example. The quality of these solvents can be determined by the absolute values of A_2, and with that, by the slopes in the plot. m-Xylene turns out to be a Θ-solvent with a slope of zero. Dioxane has a positive linear slope and therefore positive A_2, denoting it to be a good solvent. For the best solvent, chloroform, the concentration dependence of the osmotic pressure is even stronger, and it even

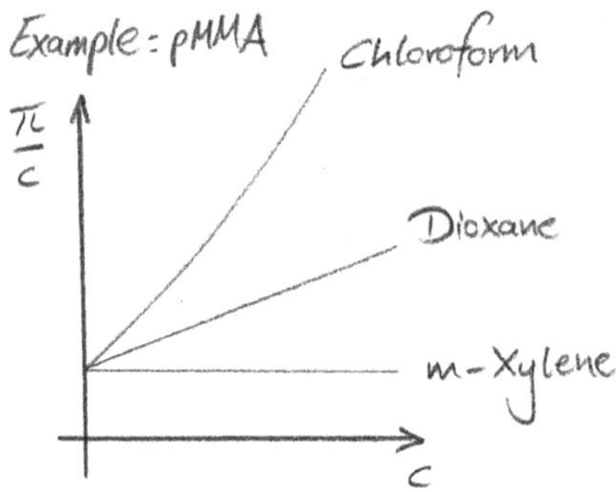

Figure 61: Schematic of the concentration-dependent reduced osmotic pressure for poly(methyl methacrylate) in different solvents. It exhibits strongly differing slopes, corresponding to strongly differing second virial coefficients, which is due to the different solvent qualities for the polymer. Picture inspired by B. Vollmert: *Grundriss der Makromolekularen Chemie*, Springer, **1962**.

51 Again, we can make an analogy to a real gas, whose equation of state can be written by a virial series as $p = \frac{RT}{V}\left(1 + \frac{B}{V} + \ldots\right)$ with the second virial coefficient $B = b - \frac{a}{RT}$, wherein a and b are the van der Waals coefficients, with a accounting for attractive intermolecular interactions, and b accounting for the covolume and other repulsive interactions of the gas molecules. At the *Boyle temperature*, attractive and repulsive interactions are at balance, causing B to be zero and the virial series thereby to simplify to the ideal gas law. The real gas then displays pseudo-ideal behavior, analogous to polymers in a solvent at the Θ-temperature.

deviates from the linear form, which makes a third nonlinear virial term necessary to fit the experimental data. Despite all these different slopes, however, the intercept is the same for all samples, as this is only dependent on the molar mass average, M_n, but not on the solvent quality.

4.3.1 Connection of the second virial coefficient, A_2, to the Flory–Huggins parameter, χ, and the excluded volume, v_e

We have already hinted at the close interrelation between the second virial coefficient A_2 to the Flory–Huggins interaction parameter χ, as both are a measure of polymer–solvent interactions. The following paragraph will examine this relationship and determine it quantitatively.

As a starting point, we take the thermodynamic definition of the osmotic pressure:

$$\Pi \equiv \left(- \frac{\partial \Delta F_{\text{mix}}}{\partial V} \right)_{n_A} = \left(- \frac{n \cdot \partial \Delta \bar{F}_{\text{mix}}}{\partial V} \right)_{n_A} \tag{4.38}$$

We have seen in the last section that the second virial coefficient A_2 is related to the reduced osmotic pressure, Π/c. Hence, we must transform the above equation into a practical expression for the osmotic pressure as a function of concentration. To achieve that, we use the expression for ΔF_{mix} that we have appraised with the lattice model of the Flory–Huggins mean-field theory in Section 4.1. This expression uses the volume fraction, ϕ, as a measure of concentration, which is connected to the total volume of the system, V, by

$$\phi_A = \frac{n_A N_A v_0}{V} \tag{4.39a}$$

Here, n_A is the number of A-molecules on the lattice, N_A the number of lattice sites per A-molecule, which is equal to the degree of polymerization of species A, and v_0 the volume of a single lattice site.

Rearranging eq. (4.39a) and expressing it in a differential form yields:

$$\partial V = \frac{n_A N_A v_0}{\partial \phi} \tag{4.39b}$$

We can insert this relation into the above expression for the osmotic pressure, and thus, change the variable V to the variable ϕ:

$$\Pi \equiv \left(- \frac{n \cdot \partial \Delta \bar{F}_{\text{mix}}}{\partial V} \right)_{n_A} = \left(\frac{\phi^2}{n_A N_A v_0} \cdot \frac{\partial \left(n_A N_A \cdot {}^{\Delta \bar{F}_{\text{mix}}} / \phi \right)}{\partial \phi} \right)_{n_A} = \left(\frac{\phi^2}{v_0} \cdot \frac{\partial \left({}^{\Delta \bar{F}}{}_{\text{mix}} / \phi \right)}{\partial \phi} \right)_{n_A} \tag{4.40}$$

The Flory–Huggins mean-field solution for the Free Energy of mixing is

$$\Delta\bar{F}_{\text{mix}} = k_B T\left(\frac{\phi}{N_A}\ln\phi + \frac{(1-\phi)}{N_B}\ln(1-\phi) + \chi\phi(1-\phi)\right) \tag{4.22}$$

We now have to differentiate this term. Unfortunately, this is difficult for the B component part of the equation, as our variable ϕ enters in a complicated form of ln $(1-\phi)$ here. To simplify this point of our calculation, we use a series expansion according to

$$\frac{(1-\phi)}{N_B}\ln(1-\phi) = \frac{1}{N_B}\left(-\phi + \frac{\phi^2}{2} + \frac{\phi^3}{6} + \ldots\right) \tag{4.41}$$

By plugging this into the term for ΔF_{mix}, we generate

$$\Delta\bar{F}_{\text{mix}} = k_B T\left(\frac{\phi}{N_A}\ln\phi + \phi\left(\chi - \frac{1}{N_B}\right) + \frac{\phi^2}{2}\left(\frac{1}{N_B} - 2\chi\right) + \frac{\phi^3}{6N_B} + \ldots\right) \tag{4.42}$$

This expression, in turn, can be inserted into our original formula for the osmotic pressure, eq. (4.40)

$$\Pi = \frac{k_B T}{v_0}\left(\frac{\phi}{N_A} + \frac{\phi^2}{2}\left(\frac{1}{N_B} - 2\chi\right) + \frac{\phi^3}{3N_B} + \ldots\right) \tag{4.43}$$

In the next step, we transform further by inserting the concentration as a variable according to $c = \phi/v_0$. This variable has the unit L^{-1}; it expresses the number of chain segments per volume, that is, per liter:

$$\Pi = k_B T\left(\frac{c}{N_A} + \frac{c^2}{2}v_0\left(\frac{1}{N_B} - 2\chi\right) + \ldots\right) \tag{4.44a}$$

Division of this formula by c generates the needed expression for the reduced osmotic pressure, Π/c

$$\Pi/c = k_B T\left(\frac{1}{N_A} + \frac{c}{2}v_0\left(\frac{1}{N_B} - 2\chi\right) + \ldots\right) \tag{4.44b}$$

When comparing the general form of the above formula to the one of eq. (4.37), we realize that they both are of kind

$$\Pi/c = \text{Thermal Energy}\left(\frac{1}{\text{Chain Length}} + \text{Interaction Parameter}\cdot c\right) \tag{4.45}$$

Equation (4.44b) has a dimensionless first summand in the parentheses; this means that the second summand must be dimensionless, too. As a result, the interaction parameter in it must have a unit of liters, L. Hence, this parameter must be some kind of volume that quantifies the polymer–solvent interactions. A quantity doing this is the excluded volume, v_e, which therefore serves as the second virial coefficient in eq. (4.44b):

$$v_e = v_0\left(\frac{1}{N_B} - 2\chi\right) \tag{4.46a}$$

Here, N_B is the degree of polymerization of the second component, which is either a solvent or a second polymer. In a polymer solution, we have $N_B = 1$. When we now plug in the Flory–Huggins parameter for the Θ-state, $\chi = 1/2$, then we end up with an excluded volume of $v_e = 0$, exactly as the definition of the Θ-state demands. For the case of an athermal solvent, by contrast, $\chi = 0$, and as a result, $v_e = v_0$. The excluded volume is then maximally positive and coincides with the monomer-segmental volume, as we have already seen in Section 3.2. In a mixture of two different polymers, a polymer blend, we have $N_B >> 1$. A Flory–Huggins parameter close to zero, $\chi \approx 0$, which is needed to successfully blend two polymers, now leads to an excluded volume also close to zero, $v_e = v_0/N_B \approx 0$. This shows us that, even though the A–B interaction in a polymer blend is very good (it must be so to mix!), the chains adopt their ideal conformation, as they do in the Θ-state.

We can further express the above equation as

$$v_e = \frac{v_0}{N_B} - 2v_0\chi = \frac{l^3}{N_B} - 2l^3\chi \tag{4.46b}$$

Here, l^3/N_B is the volume of the polymer segments (normalized to the degree of polymerization of the solvent), which corresponds to the hard-sphere branch of the Mayer *f*-function. $2l^3\chi$ is a measure of the effective segment–segment attraction, which corresponds to the potential-well part of the Mayer *f*-function. The concert of both determines the excluded volume (remember that this volume is in fact calculated from the negative integral over the Mayer *f*-function). We now see once more that for the Θ-state ($v_e = 0$), coil expansion due to the co-volume of the segments just exactly balances the coil contraction due to the net-attractive segment–segment interaction.

So far, we have expressed the reduced osmotic pressure on the molecular level, meaning we use k_BT for the thermal energy and a concentration $c = \phi/v_0$ with the unit L^{-1} to account for the number of molecules per volume. To make this more practical, we can transform eq. (4.44b) to molar values by using RT for the thermal energy and $c = \phi/v_{\text{specific,polymer}}$ with the unit $g \cdot L^{-1}$ as the concentration. For a polymer solution ($N_B = 1$), this yields

$$\Pi/c = RT\left(\frac{1}{M_n} + \frac{v^2_{\text{specific, polymer}}}{\bar{V}_{\text{Solvent}}}\left(\frac{1}{2} - \chi\right)c + \ldots\right) \tag{4.47}$$

When we compare this to the Van't Hoff equation

$$\Pi/c = RT\left(\frac{1}{M_n} + A_2c + \ldots\right) \tag{4.37}$$

we immediately recognize the connection between the second virial coefficient, A_2, and the Flory–Huggins parameter, χ:

$$A_2 = \frac{v^2_{\text{specific, polymer}}}{\bar{V}_{\text{Solvent}}}\left(\frac{1}{2} - \chi\right) \tag{4.48}$$

It is thanks to this relation that we can determine χ through an osmotic pressure experiment as described at the beginning of this chapter.

5 Mechanics and rheology of polymer systems

LESSON 10: RHEOLOGY

The most relevant properties of polymers, on which ground they have taken dominance in our materials world, is their mechanical ones. In this lesson, you will be introduced to the fundamentals of a methodology to assess these mechanics: rheology. You will become acquainted with some elementary physical and working principles in this field and get to know the most relevant material property of polymers: the elastic modulus.

5.1 Fundamentals of rheology

At the very beginning of this book, it has been pointed out that the prime goal of polymer physical chemistry is the understanding of the relations between the **structure** and **properties** of polymer-based materials, thereby bridging the fields of **polymer chemistry** and **polymer engineering**, as shown in Figure 1. To achieve this goal, so far, we have studied the structure and dynamics of single chains and the thermodynamics of multichain systems in the preceding chapters; with that, we have erected three fundamental pillars of that foresaid bridge. Now, the construction of the actual bridge will be the goal of the forthcoming chapter. This will help us to rationally and quantitatively understand why many of the polymeric materials that we encounter in our everyday lives behave the way they do. Our prime focus will be on those properties of polymers that have made greatest impact on our lives in the past decades, which is their *mechanical* properties. Hence, we will first deal with a fundamental introduction to the field of **rheology**.

The term rheology derives from the Greek words *rheos*, meaning “flow”, and *logos*, meaning “understanding”. The name reminds of a famous aphorism voiced by the Greek philosopher Simplicus, who had stated that *panta rhei*, “everything flows”. As such, rheology is the study of the flow of matter, and therefore the branch of physics that deals with the deformation and flow of materials, both solids and liquids.

5.1.1 Elementary cases of mechanical response

Two elementary extreme cases can be distinguished from one another in view of a material’s mechanical properties: it can behave either as an **elastic solid**, like a rubber band, or as a **viscous fluid**, like water or syrup.

In the first case of an ideal elastic solid, the energy introduced into the material through exertion of an external force, f, which can also be expressed in a normalized form as a *stress*, $\sigma = f/A$ (with A being the cross-sectional area of the material specimen onto which the force acts), is stored elastically. In this case, the material is

https://doi.org/10.1515/9783110672817-005

immediately deformed to an extent proportional to the applied stress, or vice versa, a counteracting stress builds up in it upon and proportional to deformation by a given extent. This proportionality of the extent of deformation, or **strain**, ε, to the **stress**, σ, is captured by **Hooke's law**:

$$\sigma = E \cdot \varepsilon \tag{5.1}$$

In this law, the strain is the relative extent of deformation, $\varepsilon = \Delta L / L_0$, that is, the change of the material specimen length, ΔL, relative to its length before deformation, L_0. The constant of proportionality is the **elastic modulus**, also referred to as **Young's modulus**, E. In a mathematical view, from eq. (5.1), we realize that the elastic or Young's modulus has a unit of $N \cdot m^{-2}$ = Pa, because the stress is described by the unit $N \cdot m^{-2}$ = Pa as well, whereas the strain, $\varepsilon = \Delta L / L_0$, is a dimensionless quantity. In a conceptual view, the modulus is an elementary material parameter that quantifies how much energy can be stored in the material upon deformation by a given extent. In this view, it is illustrative to note that the unit $N \cdot m^{-2}$ may also be written as $J \cdot m^{-3}$, which denotes the energy density in the material that we need to work against when we want to deform it; it originates from the energies and sizes of the material's building blocks (e.g., polymer-network strands of energy $k_B T$ and size ξ in a rubber). As we already see from the latter note, the capability for energy storage is directly related to the internal microscopic structure and dynamics of the material. Hence, the modulus is nothing less than a quantitative expression of a structure–property relation. With knowledge of the modulus, we can calculate from eq. (5.1) how hard it is to achieve a certain extent of deformation upon application of a given stress, which is of direct relevance for applications. As soon as the stress cedes, the stored energy is released, and the material snaps back to its original shape. This means that elastic deformation is reversible. This type of mechanical response can be visualized illustratively in the form of an elastic spring, as shown in Figure 62(A).

In the second case of an ideal viscous fluid, the exertion of an external stress also causes a deformation. Here, however, the *stress*, σ, is proportional not to the deformation itself but to its *rate*, $d\varepsilon/dt$. It follows that the material will reside in its deformed state as soon as the stress cedes. There is no energy storage in this case; instead, the energy of deformation is dissipated to the environment as heat, which means that viscous deformation is irreversible. The proportionality between the *rate of deformation*, $d\varepsilon/dt$, and the *stress*, σ, is captured by **Newton's law**:

$$\sigma = \eta \cdot \frac{d\varepsilon}{dt} = \eta \cdot \dot{\varepsilon} \tag{5.2}$$

In this law, the constant of proportionality is the **viscosity**, η. Just like the modulus in the elastic case above, the viscosity is an elementary material parameter

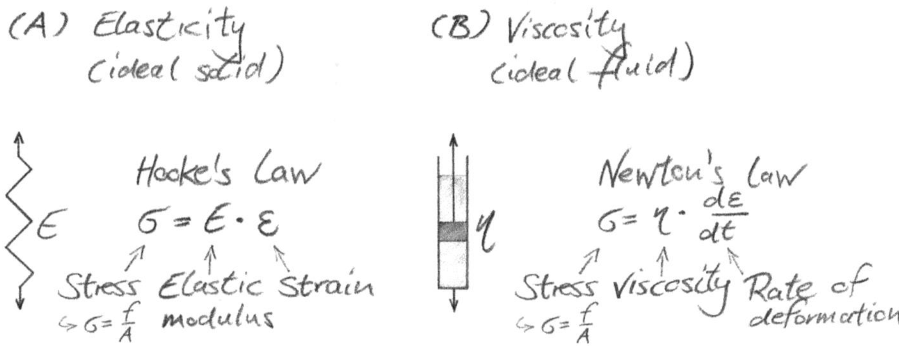

Figure 62: Elementary cases of mechanical response in rheology. **(A)** *Hooke's law*, relating the stress, σ, imposed on an *ideal elastic solid*, to its strain, ε, via the Young's modulus, E. This model can be visualized in the form of an elastic spring with a spring constant of E. **(B)** *Newton's law*, relating the stress, σ, imposed on an *ideal viscous fluid* to its time-dependent *rate* of deformation, $d\varepsilon/dt$, via the viscosity, η. This model can be visualized in the form of a dashpot filled with a fluid of viscosity η.

that quantifies how much resistance a material exerts on flow.[52] As we will see later in this chapter, the capability for exerting resistance on flow can be directly related to the internal microscopic structure and dynamics of a material. Hence, just like the elastic modulus, the viscosity is a quantitative expression of a structure–property relation. The viscous type of mechanical response can be visualized illustratively in the form of a dashpot filled with a fluid, as illustrated in Figure 2(B). In view of eq. (5.2), we realize that the viscosity has the unit Pa·s, because the stress is described by the unit $N \cdot m^{-2} = Pa$, and the rate of deformation has the unit s^{-1}.

5.1.2 Different types of deformation in rheology

Strain can be applied to a material in many directions, but we may simplify our view to two particularly relevant cases. To visualize these, consider a cube with edges A, B, and C. In this cube, A is the edge in and out-of-plane, B is the vertical edge, and C the horizontal one (see Figure 63).

In the first case, a uniform strain exerted onto the A–B face is called **uniaxial strain**. This strain, ε, can be calculated from the ratio between the deformation, ΔC, and the original length of the edge C. For the two elementary cases of an elastic solid and a viscous fluid, Hooke's and Newton's law can be applied as stated in Section 5.1.1 to calculate the stress, σ, resulting from such a uniaxial strain (see Figure 63(A)).

52 Later on, in Sections 5.6.1 and 5.7.1, we will see that the elastic modulus and the viscosity are connected to one another by one of the most central equations in the field of rheology: $\eta = E\,\tau$.

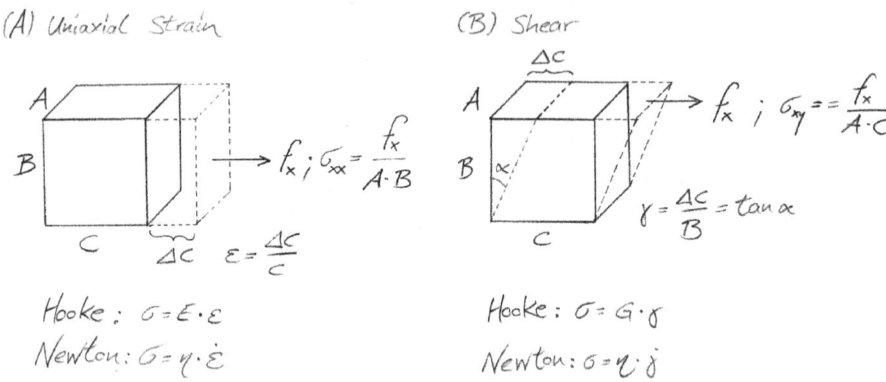

Figure 63: Two ways to deform a material: **(A)** a uniform strain exerted onto the A–B face is called *uniaxial strain*. **(B)** A strain exerted onto the A–C face is called *shear*. In both cases, the *stress*, σ, can be calculated with Hooke's law for an ideal elastic material or with Newton's law for an ideal viscous fluid.

In the second case, the stress is exerted onto the A–C face. This situation is called **shear**. The *shear strain*, γ, is calculated from the ratio between the deformation, Δc, and the original length of edge B, which corresponds to the tangent of the shearing angle α (see Figure 63(B)). For an ideal viscous fluid, the stress, σ, is still calculated using Newton's law as stated in Section 5.1.1. Hooke's law, however, needs to be modified to incorporate a new proportionality constant: the **shear modulus**, G.

Both the *Young's modulus*, E, and the *shear modulus*, G, are connected to one another in the form of

$$E = 2G(1 + \mu) \tag{5.3}$$

In eq. (5.3.), μ is named Poisson's ratio; it is a measure of the Poisson effect, a phenomenon according to which a material tends to expand in directions perpendicular to the direction of compression or, vice versa, tends to laterally contract upon extension; both is a consequence of the retention of the material body's volume upon compression or extension. μ quantifies this effect by relating the relative lateral contraction of a material specimen, expressed by the change of its thickness relative to the one before deformation, $\Delta d/d_0$, to its axial strain, $\Delta L/L_0$:

$$\mu = \frac{\Delta d/d_0}{\Delta L/L_0} \tag{5.4}$$

Incompressible objects such as fluids have a Poisson's ratio of $\mu = 0.5$. This value also commonly applies to polymers, because linear extension of a polymer sample is usually accompanied by a concurrent lateral contraction, meaning that it does not undergo a volume change upon stretching. In this case, $E = 3G$.

5.1.3 The tensor form of Hooke's law

The equations for Hooke's and Newton's laws named above are actually simplified forms for a stress acting *normal* to the sample and *in one direction only*. In their generalized forms, they are tensor equations that account for all possible spatial directions. In this generalized form, Hooke's law reads

$$\overline{\overline{\tau}} = \overline{\overline{\overline{\overline{c}}}} \cdot \overline{\overline{\gamma}} \tag{5.5}$$

with $\overline{\overline{\tau}}$ the stress tensor, $\overline{\overline{\overline{\overline{c}}}}$ the elasticity tensor, and $\overline{\overline{\gamma}}$ the strain tensor. With this nomenclature, we follow a common convention, in which – unfortunately – the tensor form of the stress is abbreviated with the symbol τ. This is inconvenient, because in polymer physics, τ is very commonly used to abbreviate the relaxation time as well.[53] To avoid this confusion, we use the symbol σ for the special-case variant of normal stress in this textbook; this is the stress with both its directional indices being the same, that is $\sigma_{11} = \tau_{11}$, $\sigma_{22} = \tau_{22}$, and $\sigma_{33} = \tau_{33}$.

To illustrate the relation of eq. (5.5), again consider a cube representing a volume section of an elastic body. If we impose a stress τ_{ik} in the direction k, the cube will deform, but now, the resulting strain will not be limited to just one spatial direction, but instead, it will have contributions for all spatial directions (see Figure 64).

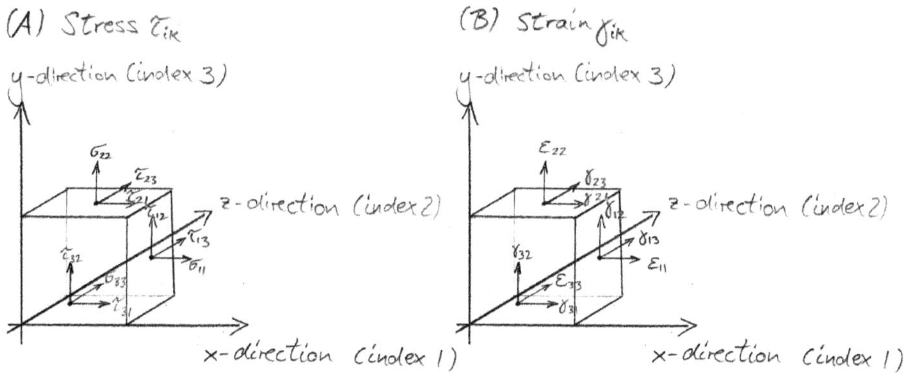

Figure 64: Deformation of an elastic body taking into account all spatial directions. **(A)** A stress τ_{ik} imposed in the direction k results in a **(B)** corresponding strain γ_{ik} in the same direction. Picture redrawn from C. Wrana: *Polymerphysik*, Springer Verlag Berlin, Heidelberg, **2014**.

53 If you are unsure whether a symbol τ in a textbook or a paper represents stress or (relaxation) time, check its physical unit: if it is Pascal, then τ represents a stress, and if it is seconds, then τ represents a time.

This relation is reflected by the following equation:

$$\tau_{ik} = \sum_{l=1}^{3}\sum_{m=1}^{3} c_{iklm} \cdot \gamma_{lm} = c_{ik11} \cdot \varepsilon_{11} + c_{ik12} \cdot \gamma_{12} + c_{ik13} \cdot \gamma_{13} +$$
$$c_{ik21} \cdot \gamma_{21} + c_{ik22} \cdot \varepsilon_{22} + c_{ik23} \cdot \gamma_{23} + c_{ik31} \cdot \gamma_{31} + c_{ik32} \cdot \gamma_{32} + c_{ik33} \cdot \varepsilon_{33} \quad (5.6)$$

The strength of each contribution is given by the elastic constant c_{iklm}. Spatial directions for each value are indicated by the indices l and m.

For a comprehensive consideration, we would have to evaluate all *nine* stress components $(\sigma_{11}, \tau_{12}, \tau_{13}, \tau_{21}, \sigma_{22}, \tau_{23}, \tau_{31}, \tau_{32}, \sigma_{33})$, which are connected by *nine* elastic constants c_{iklm} to *nine* strain components $(\varepsilon_{11}, \gamma_{12}, \gamma_{13}, \gamma_{21}, \varepsilon_{22}, \gamma_{23}, \gamma_{31}, \gamma_{32}, \varepsilon_{33})$. This adds up to 81(!) single components to fully describe the deformation of a cubic volume. Fortunately, if we consider the symmetry of the body, this expression can be severely simplified. If the body is *isotropic*, meaning it is uniform in all orientations, the number of independent constants is reduced to *two*. Polymers exhibit homogeneous and amorphous structures and can therefore be considered to be isotropic. We can now apply this to the two cases of uniaxial strain and shear deformation, which leads to the following equations for Hooke's law considering all three spatial direction x, y, and z:

Uniaxial strain:	Shear deformation:
$\sigma_{xx} = E \cdot \varepsilon_{xx};\quad \sigma_{yy} = E \cdot \varepsilon_{yy};\quad \sigma_{zz} = E \cdot \varepsilon_{zz}$	$\tau_{xy} = G \cdot \gamma_{xy};\quad \tau_{xz} = G \cdot \gamma_{xz};\quad \tau_{yz} = G \cdot \gamma_{yz}$

These are just the equations we introduced in Section 5.1.1. We can see that ε and γ are, in fact, special-case variants of $\bar{\bar{\gamma}}$, σ and τ are special-case variants of $\bar{\bar{\tau}}$, and E and G are special-case variants of $\bar{\bar{\bar{c}}}$.

5.2 Viscoelasticity

Lesson 11: Viscoelasticity

The rich mechanical properties of polymers emerge because they are neither a true fluid nor a true solid; instead, they display characteristics of both, referred to as viscoelasticity. This lesson will show how this duality manifests itself in a polymer's time- or frequency-dependent stress relaxation or strain creep. It will also show how this is captured by complex-number material constants, such as the complex modulus.

Often, materials exhibit both elastic as well as viscous properties. These materials are named **viscoelastic**. Strictly speaking, all matter exhibits this dualism. For example, imagine watching a glacier for one day. You will not be able to observe any perceivable movement of the ice and could therefore say that it is a solid. Now imagine visiting the same glacier once a day over the course of an entire year. Every time you visit, you take a photograph, and in the end you edit all pictures together to create a time-lapse movie. The glacier will appear to be flowing like a fluid. Another example is the big, ornamental glass windows that can be found in old cathedrals: often, they are thicker at their bottom than at the top. This difference is mainly related to the process with which they are produced: in historic times, church windows were manufactured upright, which meant that the cooling glass melt would flow a little from top to bottom inside the manufacturing frame. In addition to this manufacture-related thickness gradient, however, the glass of these windows still flows today, and over very long timescales (longer than centuries in fact), it causes further thickening of the window bottoms compared to their tops.

Viscoelasticity is a phenomenon most strikingly encountered in polymer systems. In contrast to glaciers and glass windows, polymer systems usually transition from a regime where their elastic properties dominate to one where the viscous properties are dominant on timescales that are right in the practically and experimentally relevant window of milliseconds to seconds, depending on the specific system at hand. This makes polymers highly relevant for practical applications, because their mechanical properties are versatile, dynamic, and, with the proper understanding of the underlying concepts that govern their mechanics, can be rationally designed to serve a specific purpose.

Polymer's mechanical properties are usually assessed and quantified in three types of rheological experiments, as will be discussed in the following subsection.

5.2.1 Elementary types of rheological experiments

5.2.1.1 The relaxation experiment

In a **relaxation experiment**, uniaxial or shear strain, ε or γ, is applied to a sample and the resulting stress in the material, σ, is recorded as a function of time, as shown in Figure 65. An elastic sample shows direct proportionality of both quantities according to Hooke's law. The proportionality factor is the time-dependent **relaxation modulus** $E(t)$. As long as strain is applied, an equivalent stress is recorded. An ideal viscous sample, instead, displays two spikes of the stress, one at the moment where the stress is abruptly increased and one at the moment where the stress is abruptly decreased. This is because according to Newton's law, only the time-dependent rate of deformation is considered, and hence, at times where the strain is constant, no stress is recorded, whereas an infinite spike of stress occurs during an abrupt change of the strain. In the case of a viscoelastic sample, the initial response is similar to the elastic one: the stress increases proportionally to the strain. Once the strain is kept at a constant level, though, some of the stored energy of deformation is dissipated as heat, leading to **stress relaxation** over time. If the sample is a viscoelastic solid, the stress will reach a plateau value with a **plateau modulus** $E_{eq} = E(t \to \infty)$.

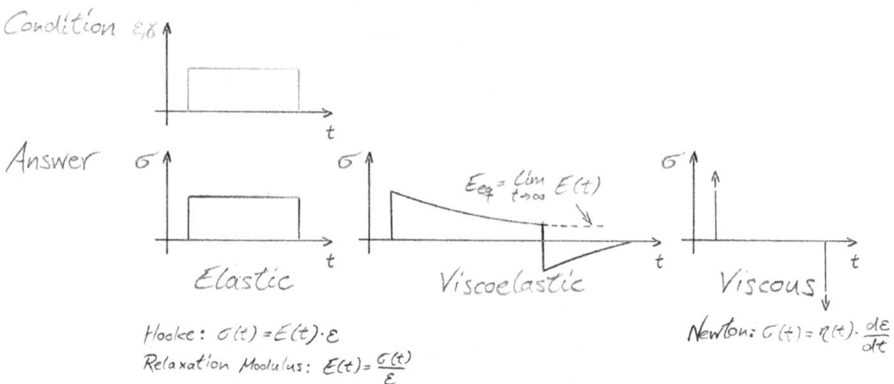

Figure 65: In a relaxation experiment, *strain*, ε or γ, is applied to a sample, and the time-dependent *stress*, σ, that builds up in the sample is recorded. An ideal elastic sample shows direct proportionality of both quantities according to Hooke's law. In an ideal viscous sample, the stress displays spikes at times of abrupt change of the strain. In a viscoelastic sample, the stress in the material relaxes to a plateau value for $t \to \infty$, because some of the energy of deformation is dissipated to heat over time.

5.2.1.2 The creep test

In a **creep test**, stress, σ, is applied to a sample, and the resulting uniaxial or shear strain of the specimen, ε or γ, is recorded as a function of time (see Figure 66). An

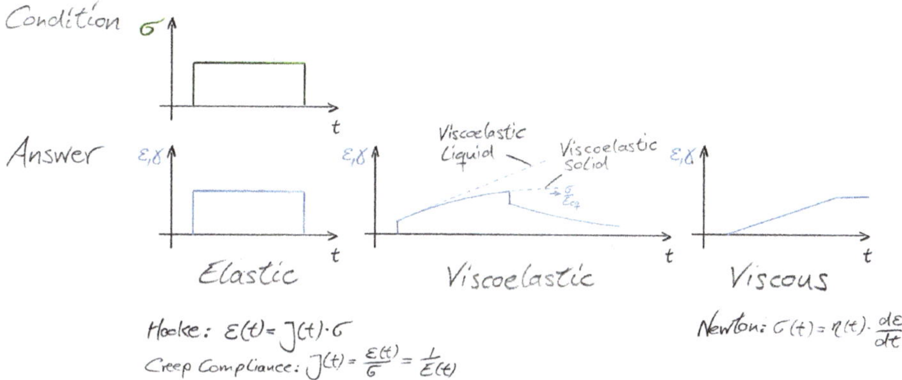

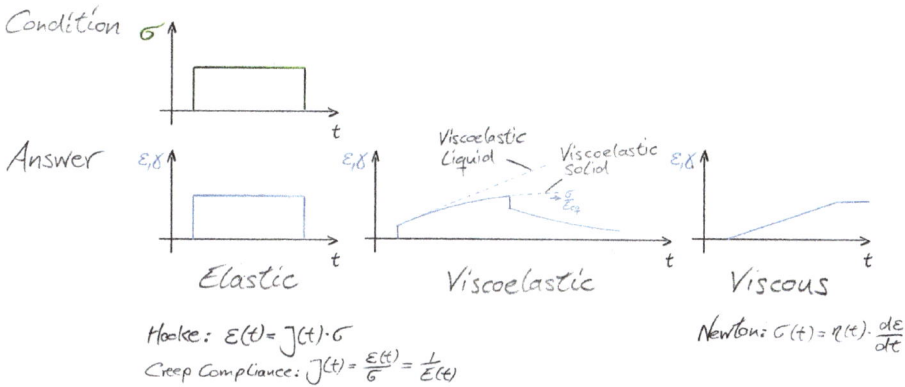

Figure 66: In a creep test, *stress*, σ, is applied to a sample, and the resulting time-dependent *strain* of it, ε or γ, is recorded. An ideal elastic sample shows direct proportionality of both quantities according to Hooke's law. In an ideal viscous sample, the strain increases linearly with time as long as constant stress is applied; the material exhibits *creep*. In a viscoelastic sample, the response is a mix of both: the recorded strain value rises instantly, but not to same extent as in the elastic sample, and with time it rises further due to creep of the material. A *viscoelastic solid* reaches a plateau for $t \to \infty$, whereas a *viscoelastic liquid* proceeds creeping steadily. Once the application of stress is released, the sample creeps back with time.

elastic sample shows direct proportionality of both quantities according to Hooke's law. The proportionality factor is called the time-dependent **creep compliance**, $J(t)$. As long as stress is applied, an equivalent strain is recorded. An ideal viscous sample exhibits a strain that increases linearly as long as the stress is kept constant; the material exhibits **creep**. Once the test is finished by release of the stress, the strain does not return its original level, thereby showcasing the irreversibility of viscous flow. In case of a viscoelastic sample, the response is a mix of both: once a constant amount of stress is applied, the recorded strain value rises instantly, but not to the same extent as in the elastic sample. With time, however, it rises further due to creep of the material. If the sample has more elastic than viscous character, it would reach a plateau for $t \to \infty$, whereas it would proceed creeping steadily if the sample has more viscous than elastic character. Once the test is finished and the application of stress is released, the sample exhibits creep in the direction back.

5.2.1.3 The dynamic experiment

Another method to probe a material's mechanical properties is the **dynamic experiment**. Here, a sinusoidally modulated stress, $\sigma = \sigma_0 \exp(i\omega t)$, is applied to the sample, and the time-dependent strain, ε or γ, is recorded, as illustrated in Figure 67. This can also be done vice versa, that is, a sinusoidally modulated strain is applied and the emerging time-dependent stress in the material is recorded. In the case of an

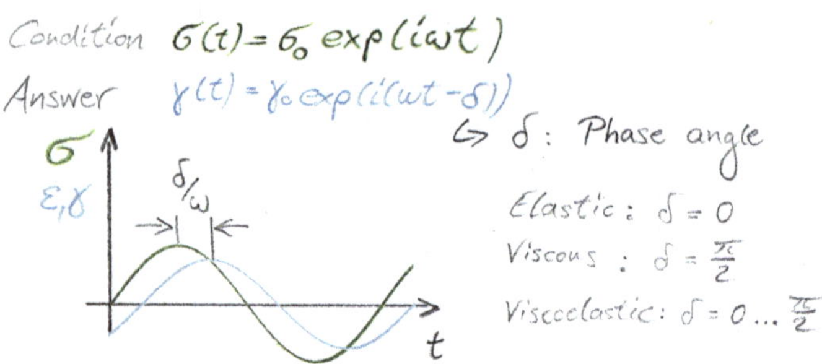

Figure 67: In a dynamic experiment, a *sinusoidal stress*, σ, is applied to a sample, and the time-dependent *strain*, ε or γ, is recorded. Both the stress and strain curves are proportional and in phase for an ideal elastic sample. By contrast, the time-dependent strain curve of an ideal viscous sample exhibits a phase shift to the stress curve with a phase angle, δ, of 90° or $\pi/2$. In the intermediate case of a viscoelastic sample, the stress and strain curves display a phase angle between 0 and $\pi/2$. In this case, δ reflects the individual contribution of the elastic and viscous parts of the sample.

ideal elastic sample, both stress and strain are in phase, as both quantities are directly proportional to each other at each time according to Hooke's law; as a result, their phase angle, δ, is zero. In the case of an ideal viscous sample, according to Newton's law, it is not the strain itself but its derivative that is proportional to the applied stress; hence, the original sinewave condition causes a cosine answer. As a result, stress and strain have a phase shift of 90°; this means that the phase angle, δ, is $\pi/2$. A viscoelastic sample will be intermediate between these two extreme cases and display a phase shift with phase angle, δ, in the range $0 < \delta < \pi/2$. From δ, the individual contribution of the elastic and viscous parts of the viscoelastic sample can be calculated. The closer δ is to zero, the more does the elastic contribution dominate, whereas the closer δ is to $\pi/2$, the more does the viscous contribution dominate.

5.3 Complex moduli

In Hooke's law, eq. (5.1), the elastic modulus reflects the ratio of stress, σ, to strain, ε or γ. In the case of a dynamic experiment as just discussed, this modulus is a complex quantity, $E^\star$ or $G^\star$, connecting the ratio of the sinusoidally modulated stress and strain and the phase angle between them, δ, as follows:

$$E^\star = \frac{\sigma}{\varepsilon} = \frac{\sigma_0 \exp(i\omega t)}{\varepsilon_0 \exp(i(\omega t - \delta))} = \frac{\sigma_0}{\varepsilon_0} \exp(i\delta) \tag{5.7a}$$

$$G^\star = \frac{\sigma}{\gamma} = \frac{\sigma_0 \exp(i\omega t)}{\gamma_0 \exp(i(\omega t - \delta))} = \frac{\sigma_0}{\gamma_0} \exp(i\delta) \tag{5.7b}$$

With Euler's formula, this can be rewritten as a trigonometric function with a real and an imaginary part; these two parts can be viewed to be two different parts of the **complex modulus:**

$$\begin{aligned} E^{\star} &= \frac{\sigma_0}{\varepsilon_0}(\cos\delta + i\sin\delta) \\ &= E' + iE'' \end{aligned} \tag{5.8a}$$

$$\begin{aligned} G^{\star} &= \frac{\sigma_0}{\gamma_0}(\cos\delta + i\sin\delta) \\ &= G' + iG'' \end{aligned} \tag{5.8b}$$

The first part is the **storage modulus,** E' or G'; it captures the sample's elastic properties. The second part is the **loss modulus,** E'' or G''; it captures the sample's viscous properties.

Storage modulus:

$$E' = \frac{\sigma_0}{\varepsilon_0}\cos\delta \tag{5.9a}$$

$$G' = \frac{\sigma_0}{\gamma_0}\cos\delta \tag{5.9b}$$

Loss modulus:

$$E'' = \frac{\sigma_0}{\varepsilon_0}\sin\delta \tag{5.10a}$$

$$G'' = \frac{\sigma_0}{\gamma_0}\sin\delta \tag{5.10b}$$

Depending on the phase angle δ, the first or the second contribution will dominate the overall modulus $E^{\star}$ or $G^{\star}$. If $\delta = 0$, then E'' and G'' are zero, such that only the elastic parts E' or G' contribute to $E^{\star}$ or $G^{\star}$. By contrast, if $\delta = \pi/2$, then E' and G' are zero, such that only the viscous parts E'' and G'' contribute to $E^{\star}$ or $G^{\star}$. In between these two extremes, both parts contribute to $E^{\star}$ or $G^{\star}$, and δ determines with what relative magnitude they do.

The extent of the storage and the loss modulus contributions to $E^{\star}$ and $G^{\star}$ can be judged from their ratio E''/E' and G''/G'. Following the simple trigonometric evaluation shown in Figure 68, or alternatively, the simple trigonometric identities of eqs. (5.11), these ratios can be described as the tangent of δ. It is called the **loss tangent**, and its value describes the contributions of the elastic and the viscous parts to a sample's viscoelastic properties in one simple number. $\tan\delta > 1$ means that the value of the loss modulus is larger than that of the storage modulus. The sample's mechanical properties are then dominated by its viscous part, and it is therefore called a *viscoelastic* *liquid.* When tan $\delta < 1$, by contrast, the sample's mechanical properties are dominated by its elastic contribution. The value of its storage modulus is then larger than that of its loss modulus, and it is therefore called a *viscoelastic solid.* When tan $\delta = 1$, both the elastic and the viscous parts contribute equally.

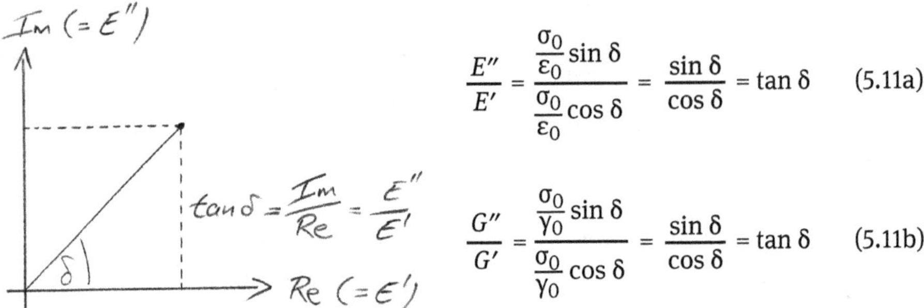

$$\frac{E''}{E'} = \frac{\frac{\sigma_0}{\varepsilon_0}\sin\delta}{\frac{\sigma_0}{\varepsilon_0}\cos\delta} = \frac{\sin\delta}{\cos\delta} = \tan\delta \tag{5.11a}$$

$$\frac{G''}{G'} = \frac{\frac{\sigma_0}{\gamma_0}\sin\delta}{\frac{\sigma_0}{\gamma_0}\cos\delta} = \frac{\sin\delta}{\cos\delta} = \tan\delta \tag{5.11b}$$

Figure 68: Visual representation of the complex modulus E^* in an Argand diagram. The abscissa represents the real part of the complex modulus, E', whereas the ordinate represents the imaginary part, E''. A vector connecting the value of E^* and the origin exhibits a slope of tan δ, the *loss tangent*.

The contributions of the storage modulus, E' or G', and the loss modulus, E'' or G'', to the complex modulus E^* or G^* can be visualized in an experiment as shown in Figure 69. When a person drops a rubber ball from the hand to the floor, it bounces off the floor and shoots back up, but it does not reach its original height, because some of the energy is dissipated to heat during the bounce. This amount of

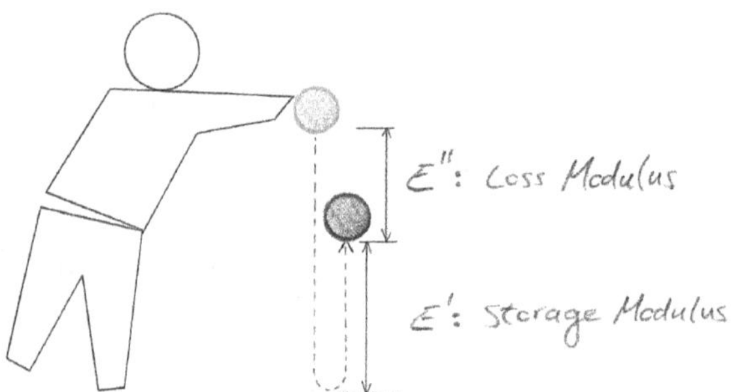

Figure 69: Visualization of the contributions of the loss modulus, E'', and the storage modulus, E', to the complex modulus E^*. In an experiment, a rubber ball is dropped, but does not bounce back to its original height. The amount of energy dissipated to heat corresponds to the loss in height after the bounce; this is the loss modulus, E''. The residual height of the bounce corresponds to the energy stored elastically within the material; this is the storage modulus, E'. An ideal elastic ball would regain its original position, whereas an ideal viscous ball would not bounce at all.

energy can be directly related to the loss in height, and this is the contribution of the loss modulus, E'' or G'', to $E^\star$ or $G^\star$. The height that the ball is able to gain back when bouncing is directly related to the energy stored elastically within the material. This is the contribution of the storage modulus, E' or G' to $E^\star$ or $G^\star$.

The latter example is a direct illustration of a structure–property relation. The extent of the rubber ball bounce is a direct resemblance of the loss tangent of its rubbery material, which is a ratio of the moduli E'' (or G'') and E' (or G'). These moduli, in turn, have to do with the material's structure. As we will see later, in Section 5.9, the ability of a rubbery polymer network to store mechanical energy is proportional to its polymer-chain crosslinking density, as this determines the tightness of the polymer-network meshes. The ability for energy dissipation, by contrast, has to do with structural motifs that are linked to the network with only one extremity, such as loops or dangling chains, as these may dissipate external deformation energy. All of the latter has to do with the network structure, and based on the aforesaid, this structure directly translates into properties. Although the example with a rubber ball may appear silly, the same holds for other rubbery products such as shoe soles. Tailoring their properties to either good damping (as requested by customers for leisure activities) or good bounce (as requested by customers for professional marathon running) can be achieved on the very same basis.

5.4 Viscous flow

In addition to the elastic Young's or shear modulus $E^\star$ or $G^\star$, other quantities that we have introduced in the previous paragraphs, such as the creep compliance, J, and the viscosity, η, can also be determined in a dynamic experiment and therefore be expressed as complex quantities.

The **complex creep compliance,** $J^\star$, which is the reciprocal of the complex modulus, $1/E^\star$ or $1/G^\star$, reflects the ratio of strain, ε or γ, to stress, σ. For the case of oscillatory shear deformation, it calculates as follows:

$$J^\star = \frac{\gamma}{\sigma} = \frac{\gamma_0}{\sigma_0}\exp(-i\delta) = \frac{1}{G^\star} = \frac{1}{G' + iG''} \tag{5.12a}$$

We can expand this expression by $\frac{G' - iG''}{G' - iG''}$ to gain

$$J^\star = \frac{G' - iG''}{(G' + iG'')(G' - iG'')} = \frac{G' - iG''}{G'^2 + G''^2} = \frac{G'}{G'^2 + G''^2} - i\frac{G''}{G'^2 + G''^2} = J' - iJ'' \tag{5.12b}$$

The complex viscosity, $\eta^\star$, is calculated analogously. It reflects the ratio of stress, σ, to the strain rate, $\mathrm{d}\varepsilon/\mathrm{d}t$ or $\mathrm{d}\gamma/\mathrm{d}t$. For an oscillatory shear deformation experiment, it calculates as follows:

$$\eta^* = \frac{\sigma}{\left(\frac{d\gamma}{dt}\right)} = \frac{\sigma_0 \exp(i\omega t)}{\frac{d}{dt}\left(\gamma_0 \exp(i(\omega t - \delta))\right)} = \frac{\sigma_0 \exp(i\omega t)}{i\omega\gamma_0\left(\exp(i(\omega t - \delta))\right)} = \frac{\sigma_0}{\gamma_0}\frac{1}{i\omega}\exp(i\delta) \tag{5.13a}$$

Insertion of $G^* = \frac{\sigma_0}{\gamma_0}\exp(i\delta)$ transforms the equation to

$$\eta^* = \frac{G^*}{i\omega} = \frac{G' + iG''}{i\omega} = \frac{G'}{i\omega} + \frac{iG''}{i\omega} \tag{5.13b}$$

By expanding the first summand with i/i and cancelling out the i in the second summand, we can write

$$\frac{iG'}{ii\omega} + \frac{G''}{\omega} = \frac{iG'}{-1\omega} + \frac{G''}{\omega} = \frac{G''}{\omega} - i\frac{G'}{\omega} = \eta' - i\eta'' \tag{5.13c}$$

We can see that the complex viscosity, η^*, is connected to the complex moduli, E^* and G^*, by the frequency, ω. It should be noted here that η' is connected to E'' and G'', whereas η'' is connected to E' and G'. This makes sense: the real part of the viscosity accounts for the dissipation of energy by a materiel, just as the imaginary part of E^* and G^* does.

This viscosity of a polymer solution is often quite different than that of a solution of small molecules that exhibits **Newtonian flow** according to eq. (5.2). The latter's viscosity is independent of the strain rate or shear rate $d\varepsilon/dt$ or $d\gamma/dt$. If we plot the stress, σ, as a function of that rate (such a plot is named *flow curve*), the viscosity is the slope of a straight line in this case, as this plot is a simple graphical representation of Newton's law. In polymer solutions and melts, however, the relation of stress to the strain rate or shear rate is often nonlinear. Instead, polymer systems often exhibit **shear-thinning**, that is, a *decrease* of the viscosity at increasing shear rate. The reason for this is because shear forces destroy possible associate structures and orient the chains along the direction of flow, thereby enabling the polymer solution or melt to flow more freely by reducing the friction and mutual impairment between their chains. Due to this structural change, shear-thinning has been termed *structural viscosity* by Wolfgang Ostwald junior in 1825.

Because their viscosity does not relate linearly to the shear rate, polymer systems are often discussed in terms of the **zero shear viscosity**, η_0, which is obtained by extrapolating the shear-thinning flow curve in Figure 70(A) to a shear rate of zero. In this limit, the shear-thinning flow curve coincides with the ideal Newtonian one. Shear-thinning properties, nevertheless, are actually wanted for many applications. For example, wall paint is specifically engineered such to be shear-thinning, because in this way, it can be applied with ease onto a wall through the exertion of shear by brushing, whereas once this is done, it will stay on the wall and dry instead of drip. Shear-thinning polymers are also used in everyday-life products such as shrink foils. These foils are produced by extrusion of a shear-thinning polymer melt in which the chains orient themselves when squeezed through the extrusion

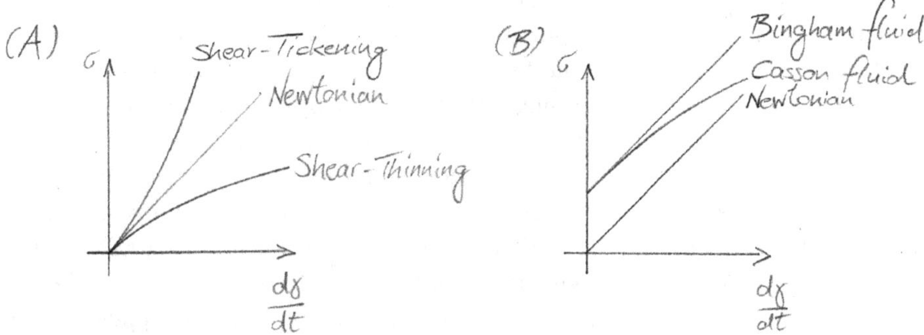

Figure 70: Plot of a *flow curve*, which is a graphical representation of stress, σ, as a function of shear rate, $d\gamma/dt$, of a flowing liquid. **(A)** In case of a Newtonian fluid, we obtain a straight line passing through the origin in which the slope is the viscosity, η, which is independent of the shear rate. This is different for the case of non-Newtonian flow, which can be observed in different manifestations. In the case of *shear-thickening*, the viscosity increases with the of shear rate, whereas in the case of *shear-thinning*, the viscosity decreases with the shear rate. The latter is often found in polymer systems, as shear breaks associates and orients polymer chains, thereby enabling them slide against each other more easily. **(B)** Further deviations from Newtonian flow are cases in which the flow curve exhibits an intercept different from zero. A *Bingham fluid* is solid at low stress, but starts to flow beyond a specific stress threshold. A *Casson fluid* is a special case of such a fluid that exhibits additional *shear-thinning* once flow sets in.

nozzle. After quick cooling and solidification, this chain orientation is frozen in the foil material. When wrapping an object with this foil and heating it, the oriented chains will relax to their original random conformation, and thus, the foil shrinks. Resolidification then locks this state. This is especially useful for materials that require an airtight packaging such as consumables.

A special case is the **Bingham fluid:** This is a material that has a flow threshold at a specific **yield stress.** It acts as a solid at zero and very low shear below the threshold, but flows like a Newtonian fluid when the shear rate is increased above it. This mechanical behavior is often attributed to secondary interactions such as Van der Waals forces or hydrogen bonds in the material, which build up a three-dimensional network structure that gives the material solid-like properties. These assemblies, however, can be broken by shear due to the low binding energies of the transient interactions involved. Bingham fluids known from everyday life include tooth paste, mayonnaise, or ketchup. Similar to that is a **Casson fluid**, which acts like a Bingham fluid but then furthermore exhibits shear-thinning. Melted chocolate is an example of a Casson fluid.

The opposite type of shear-altering behavior is called **shear-thickening**, which is a common feature of particulate suspensions, whereas it is less common for polymer solutions and melts. Here, transient hydrodynamic clusters of particles can form and break spontaneously in the material. At high shear rates, the shear oscillation period gets shorter than the transient lifetime of these clusters, such that they become permanent on the timescale of the experiment, thereby imparting hindrance on flow. A prime example is a water–sand suspension, that means wet sand. When carefully burying your feet into it on the beach, you can easily remove them from the sand when imposing only low shear by moving your feet slowly. However, if you impose high shear and try to abruptly pull your feet out of the wet sand, they will be stuck as if they were cemented in concrete.

Two further variants of complex rheological behavior are *thixotropy* and *rheopexy*. In a thixotropic case, the viscosity is observed to decrease over time during the experiment. An illustrative explanation for such a behavior is a time-dependent breakdown of a house-of-cards structure in the sample. In the opposite case, named rheopexy, the viscosity increases over time. There is no illustrative picture for that.

5.5 Methodology in rheology

Lesson 12: Practice and Theory of Rheology

So far, you got a notion about the viscoelastic nature of polymers, how it manifests itself in relaxation and creep experiments, and how it is reflected in complex material parameters. The following lesson adds information on how these parameters are actually probed in experiments, and how the phenomenology of viscoelasticity can be treated in simplifying mechanical models: the Maxwell model and the Kelvin–Voigt model. We will see how these models are conceptually very simple but yet powerful in quantitatively capturing a polymer's time- or frequency-dependent mechanics.

5.5.1 Oscillatory shear rheology

The most common and experimentally easiest way to assess the viscoelasticity of a sample is the method of **oscillatory shear rheology**. Here, a sample is placed on a lower plate that is usually an even plane, and an upper plate is lowered onto it, as shown in Figure 71. A motor then turns the upper part in an oscillatory manner, which results in a torque that is exerted by the sample-medium's resistance in between the two plates. From this torque, the values for the moduli are calculated. Sometimes, a very shallow cone is used instead of a plate as the upper geometry for the following reason: the turning motion creates a shear-force gradient facing outward from the sample's center. The cone's angled geometry eliminates this gradient and ensures a constant shear force throughout the sample. This, however, only works at a very defined gap size, which is the distance between the upper and

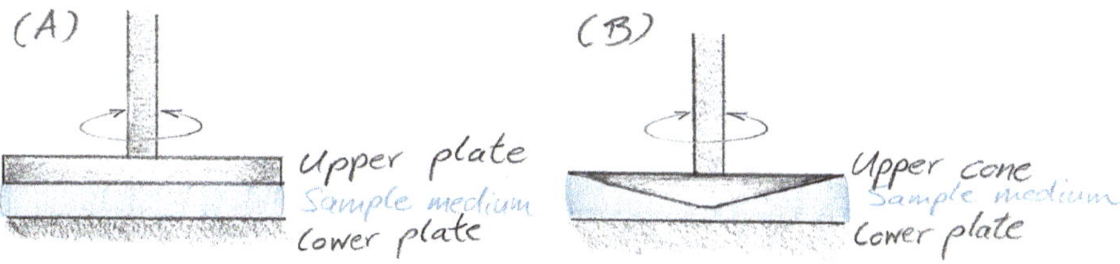

Figure 71: Schematic representation of a setup for a shear rheology experiment. The sample is applied onto the lower part of the measuring geometry, which is usually an even plane. Then, the upper part of the geometry is lowered onto the sample and sheared against the lower plate, either in a continuous or in an oscillating manner. This upper part is either also an even plate, thereby enabling the use of a user-defined gap size (**A**), or it is a shallow cone, thereby eliminating possible disruptive shear-force gradients throughout the sample, which, however, only works at a predetermined gap size (**B**).

lower geometry. Polymer solutions are therefore often measured with a cone–plate geometry, whereas polymer gels and melts are mostly measured using plate–plate geometries.

5.5.2 Microrheology

Another way to capture a sample's viscoelastic properties is the method of *microrheology*. Here, a population of nano- or micrometer-sized probes is immersed within the sample, which then explores it, as sketched in Figure 72. By quantifying the relative ease of these probes' movement through the sample medium, the sample's viscoelastic characteristics can be quantified. As the probes are small, this information can be obtained on a local microscopic scale. Furthermore, if methods of imaging microscopy are used to measure the probe motion, the information on the sample's viscoelastic mechanics is even obtained in a spatially resolved manner. With that, potential heterogeneity of the sample composition or structure can be determined.

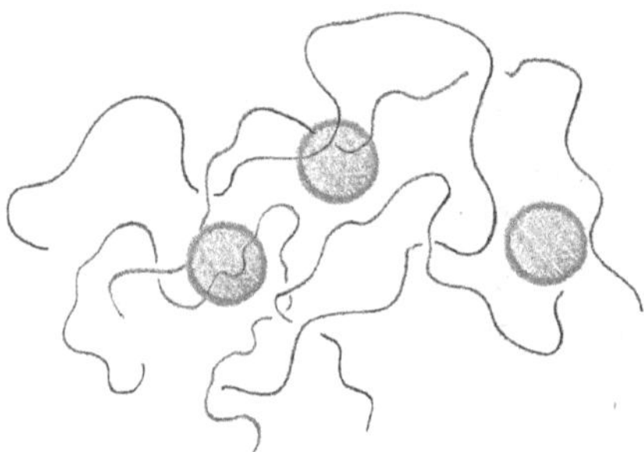

Figure 72: Schematic of a polymer sample loaded with nanoparticles that explore it by either random-thermal or externally directed motion. In the method of microrheology, observation of this probe motion is used to draw quantitative conclusions about the viscoelastic properties of the surrounding sample medium.

There are two variants of microrheology. In an *active microrheology* experiment, the probe particles are displaced by external forces. In a *passive microrheology* experiment, by contrast, only the ever-present thermal energy moves the probes.

5.5.2.1 Active microrheology

A prime example of an active microrheological method is **magnetic-bead rheometry**. Here, the sample is loaded with spherical magnetic microparticles that can be moved by application of an external magnetic field. The magnetic force, $\vec{f}_{mag}$, is related to the product of the magnetic moment induced in the bead, $\vec{M}(t)$, and the external field gradient, $\partial\vec{B}(t)/\partial x$, which can also be expressed as the magnetic susceptibility of the bead, χ, multiplied by its volume, V, and the field strength times its gradient, $\vec{B}(t)\frac{\partial \vec{B}(t)}{\partial x}$:

$$\vec{f}_{\text{mag},x}(t) = \vec{M}(t)\frac{\partial \vec{B}(t)}{\partial x} = \chi V \vec{B}(t)\frac{\partial \vec{B}(t)}{\partial x} \tag{5.14}$$

The resulting displacement of the bead can be monitored by microscopy and then translated into the viscoelastic properties in a spatially resolved manner. In the following, we first detail on how this works for the elementary case of a purely viscous medium. In such a medium, the application of a constant magnetic force, $\vec{f}_{\text{mag}}$, accelerates the probe particles, accompanied by emergence of a counteracting frictional force, $\vec{f}_{\text{frict}}$. This force is related to the particle's velocity, $\vec{v}$, and the viscosity of the medium, η, by Stoke's law. In a steady state, these two forces are at balance, and the particles reach a constant velocity. At this state, knowledge of the external magnetic force on the basis of eq. (5.14) along with knowledge of the probe particle size, r, and measurement of their velocity in the sample, $\vec{v}$, allows the viscosity, η, to be calculated:

$$\vec{f}_{\text{mag}} = \vec{f}_{\text{frict}} = 6\pi\eta r\vec{v} \tag{5.15}$$

When the sample is viscoelastic, its properties are best captured in a dynamic experiment, in which the magnetic field is sinusoidally modulated. The resulting time-dependent displacement of the bead, $x(t)$, is given by the following expression:

$$x(t) = x_0 \exp(i(\omega t - \varphi)) \tag{5.16}$$

As we have seen in Section 5.3, a complex exponential can be rewritten as a trigonometric function using Euler's formula. Analogous to the dynamic macrorheological experiment, we can find the frequency-dependent storage and loss moduli, $G'(\omega)$ and $G''(\omega)$, as follows:

$$G'(\omega) = \frac{f_0}{6\pi r|x_0\omega|}\cos\varphi \tag{5.17a}$$

$$G''(\omega) = \frac{f_0}{6\pi r|x_0\omega|}\sin\varphi \tag{5.17b}$$

5.5.2.2 Passive microrheology

Passive microrheology experiments are based on the same principles as active ones, but in contrast to those, the passive variant solely relies on thermal energy, k_BT, to drive the probe-particle motion through the sample. Based on that premise, the probes' mean-square displacement, $\Delta x^2(t)$, is observed as a function of time to characterize the sample's viscoelastic properties.

In a purely viscous medium, only Fickian diffusion of the probes takes place. In this case, the probes' mean-square displacement obeys the Einstein–Smulochowski equation:

$$\langle \Delta x^2(t) \rangle = 6Dt \tag{5.18}$$

Together with the Stokes–Einstein equation for the diffusion of spherical particles in a viscous medium, we can express the sample's viscosity as follows:

$$\eta\left(\hat{=} G''(t)\right) = \frac{k_B T}{6\pi D r} = \frac{k_B T t}{\pi r \langle \Delta x^2(t) \rangle} \tag{5.19}$$

In a purely elastic medium, the diffusion of the probes is constrained. They may move a certain short distance during which they do not yet realize the constraint exerted by the solid medium, but as soon as they feel the elastic trapping of their surrounding, they cannot move further. This means that their time-dependent mean-square displacement, $\langle \Delta x^2(t) \rangle$, first exhibits free diffusion with Einstein–Smulochowski-type scaling, but then eventually reaches a plateau value, $\langle \Delta x^2 \rangle_p$ (Figure 73). Balancing the driving thermal energy to the dragging elastic energy, with the latter expressed in a Hooke-type form with the material's spring constant, κ, yields:

$$k_B T = \frac{1}{2} \kappa \langle \Delta x^2 \rangle_p \tag{5.20}$$

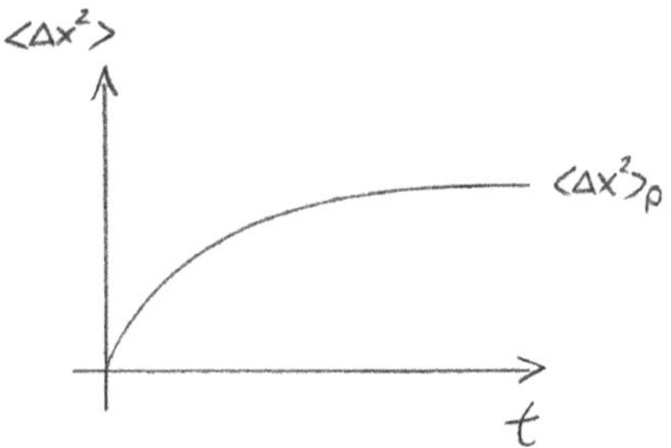

Figure 73: Mean-square displacement of probe particles, $\langle \Delta x^2(t) \rangle$, in an elastic medium as a function of time, *t*. At short times, the particles travel so short distances that they do not yet realize the elastic trapping in their surrounding. At long times, by contrast, the trapping is fully operative and prevents motion of the probes further than a distance reflected by the balance of their thermal driving energy and the counteracting elastic energy of the medium.

The loss modulus, G'', can be neglected in a purely elastic environment; as a consequence, the storage modulus, G', can be determined as

$$\Rightarrow \; G'(t) \sim \frac{\kappa}{r} \approx \frac{k_B T}{r \langle \Delta x^2 \rangle_p} \tag{5.21}$$

The intermediate case of a viscoelastic medium relates to the intermediate part of the curve in Figure 73, where the probes partially but not yet fully experience elastic constraint on their diffusive motion. In this range, the Einstein–Smulochowski equation (5.18) is extended by an exponent $\alpha(t)$:

$$\langle \Delta x^2(t) \rangle \sim t^{\alpha} \tag{5.22}$$

In a purely viscous medium, $\alpha(t) = 1$ and the probes can diffuse freely. In a purely elastic medium, by contrast, $\alpha(t) = 0$, and we find a plateau value for the probes' mean-square displacement, as described above. When $0 < \alpha(t) < 1$, the probe displacement is called *subdiffusive*. This is the case in a medium that has both viscous and elastic properties. The curve in Figure 73 shows that $\alpha(t)$ is time dependent. At short times, the potential elastic constraint of the surrounding medium is not yet felt by the moving particles, such that $\alpha(t) \approx 1$. At longer times, the particles more and more experience such constraint, such that $\alpha(t) < 1$ (more and more as time proceeds), and eventually, in the long-time limit, the elastic constraint is fully felt and prevents particle motion further than the plateau value of $\langle \Delta x^2 \rangle_p$, such that $\alpha(t) = 0$. From $\alpha(t)$, the complex modulus, $G^\star$, with its two parts G' and G'', can be determined based on a calculation by Mason and Weitz:

$$G'(\omega) = G(\omega) \cos\left[\alpha(\omega)\frac{\pi}{2}\right] \tag{5.23a}$$

$$G''(\omega) = G(\omega) \sin\left[\alpha(\omega)\frac{\pi}{2}\right] \tag{5.23b}$$

with

$$G(\omega) = \frac{k_B T}{\pi r \langle \Delta x^2 \left(\frac{1}{\omega}\right) \rangle \Gamma[1 + \alpha(\omega)]} \tag{5.23c}$$

5.6 Principles of viscoelasticity

5.6.1 Viscoelastic fluids: the Maxwell model

Now that we have understood that many materials, especially the soft-matter type, display mechanical properties with both elastic *and* viscous contributions, we have to find a way to mathematically describe this viscoelastic mechanics. A suitable approach is to start from simple mechanical models that combine both the stereotype mechanical element of an elastic body, which is a spring, and the stereotype element of a viscous fluid, which is a dashpot. The easiest way to combine these two elements is to connect them in series, as shown in Figure 74. This is done in the **Maxwell model**. When stress is applied to this model by an external deforming force, the elastic spring element

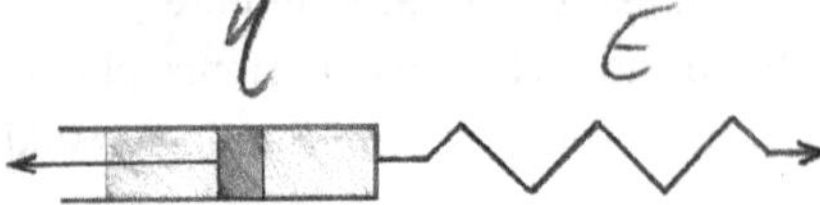

Figure 74: A *Maxwell element* is composed of a purely viscous damper (a dashpot) and a purely elastic spring connected in series. When stress is applied, the elastic spring deforms instantly, followed by delayed irreversible deformation of the viscous damper. Once the stress is released, only the spring snaps back to its original position, whereas the dashpot remains deformed.

deforms instantly, followed by a delayed irreversible deformation of the viscous damper element that relaxes the stress in the deformed spring as far as possible. If that relaxation is not allowed to come to completion, there is residual stress in the system, but once this stress is released externally, only the spring element will snap back to its original position, whereas the dashpot element will remain deformed irreversibly.

The applied stress is equal in both parts, $\sigma = \sigma_1 = \sigma_2$, whereas the total strain is the sum of the strain of the two separate parts, $\varepsilon = \varepsilon_1 + \varepsilon_2$. The same holds true for its time derivative

$$\frac{d\varepsilon}{dt} = \frac{d\varepsilon_1}{dt} + \frac{d\varepsilon_2}{dt} \tag{5.24}$$

The elastic contribution by the spring is given by Hooke's law; in differential form it reads

$$\frac{d\sigma}{dt} = E\frac{d\varepsilon_1}{dt} \tag{5.25}$$

The viscous contribution from the dashpot is described by Newton's law as

$$\sigma = \eta\frac{d\varepsilon_2}{dt} \tag{5.26}$$

We can now use the latter two equations and insert them in eq. (5.24) to generate

$$\frac{d\varepsilon}{dt} = \frac{1}{E}\frac{d\sigma}{dt} + \frac{\sigma}{\eta} \tag{5.27}$$

Solving this differential equation with the boundary condition of constant deformation, $d\varepsilon/dt = 0$, yields

$$\sigma(t) = \sigma_0 \exp\left(-\frac{Et}{\eta}\right) = \sigma_0 \exp\left(-\frac{t}{\tau}\right) \tag{5.28}$$

The latter equation quantifies the time-dependent stress relaxation of the Maxwell element; it contains a parameter that is essential for that: the **relaxation time,** $\tau = \eta/E$. This parameter connects two macroscopic quantities, the elastic modulus, E, and the viscosity, η, by a microscopic time constant, τ.[54] In a viscoelastic medium, τ is the **characteristic time of molecular rearrangement**. On timescales longer than τ, the molecules (or macromolecules in case of a polymer sample) can fully rearrange themselves, such that viscous flow dominates the viscoelastic properties. By contrast, on timescales shorter than τ, the (macro)molecules cannot yet rearrange themselves, such that elastic solidity dominates the viscoelastic properties. At the timescale of τ, the one behavior transitions into the other, and we observe marked viscoelastic properties. For polymers, τ is in the range of microseconds to seconds. With that, it falls right into the domain that many experiments and practical applications cover. As a result, polymers display rich viscoelastic mechanics in practice. The exact value of τ, and with that, the extent of viscous or elastic dominance on their mechanics, depends on parameters such as the polymer size, shape, and interactions.

From the preceding mathematical consideration, we see that the Maxwell element is good to model an experimental situation of constant strain, which leads to time-dependent stress relaxation, as shown on the left-hand side of Figure 75; this is

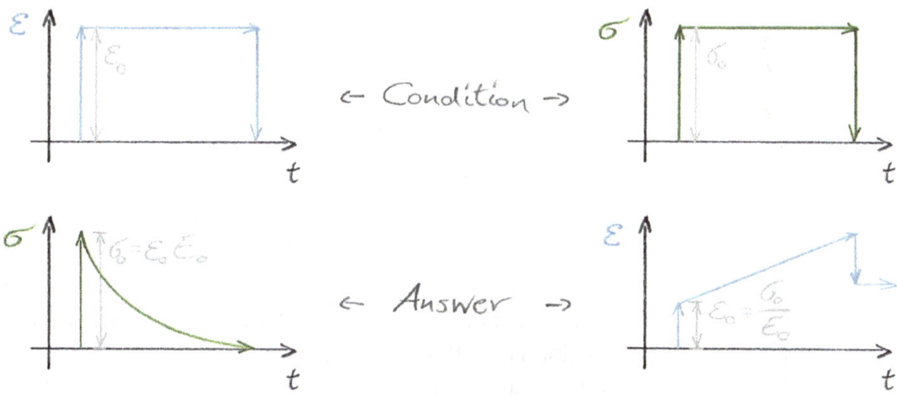

Figure 75: Relaxation experiment (left) and creep test (right) applied to a *Maxwell element*. During the relaxation experiment, the spring is first stretched, but then the damper follows and relieves the spring, such that stress can relax over time. In the creep experiment, the spring first shows an initial response that is then followed by the dashpot, which starts to exhibit normal Newtonian flow, such that the deformation steadily proceeds. Note that the latter behavior is not creep in a sense as we have defined it in Section 5.2.1.2, but instead, just flow. Hence, the Maxwell element is good at modeling stress relaxation in a viscoelastic fluid, but not creep of a viscoelastic solid.

---

54 At this point, the identity $\eta = E\,\tau$ was introduced as a pure mathematical tool to replace η/E in eq. (5.28) by a time constant τ to turn the exponential decay function $\exp(-Et/\eta)$ into the mathematically general form $\exp(-t/\tau)$. With that, we have obtained the identity $\eta = E\,\tau$ as a sort of "by product". It is, however, one of the most central equations in the field of rheology, with very fundamental physical meaning. Later on, in Section 5.7.1, we will derive it again "properly". Anyhow, we may already understand it conceptually here: viscosity (η) is composed (= a product) of the ability of a medium to initially store energy upon deformation (E) times how long it takes to dissipate that (τ).

typical for a viscoelastic fluid. By contrast, if we consider the opposite case of a constant stress, $d\sigma/dt = 0$, then eq. (5.27) just turns into Newton's law. As a result, in this case, the Maxwell element just responds with some initial strain of the spring followed by normal Newtonian flow of the dashpot, as shown on the right-hand side of Figure 75. This is not realistic; we know from Section 5.2.1.2 that a viscoelastic solid exhibits time-dependent creep under the condition of constant stress instead.

As an alternative to the purely mathematical discussion above, we can also interpret the mechanical responses of the Maxwell element shown in Figure 75 in a conceptual way. If we apply constant strain to the Maxwell element, as shown in the upper-left schematic in Figure 75, the resulting time-dependent stress initially rises sharply, but then drops back to its original value over time, as sketched in the lower left in Figure 75. This is because the spring is stretched instantly, but then the dashpot follows and relieves the strained spring. In this way, stress can relax over time, following an exponential function as in eq. (5.28):

$$\sigma(t) = \sigma_0 \exp\left(-\frac{t}{\tau}\right) = \varepsilon_0 E_0 \exp\left(-\frac{t}{\tau}\right) \tag{5.29a}$$

From that, we can calculate a time-dependent elastic modulus, E, as follows:

$$E(t) = E_0 \exp\left(-\frac{t}{\tau}\right) \tag{5.29b}$$

By contrast, if we apply constant stress to the Maxwell element, as sketched in the upper right of Figure 75, the resulting time-dependent strain initially rises to some extent, and then grows linearly as long as stress is applied, as shown in the lower right of Figure 75. Once the stress ceases, the strain drops by the exact same amount as it rose at the start, but retains the increase it has gained over time. Conceptually, the initial response can be attributed to the instant deformation of the spring. The dashpot follows and starts to flow, such that the deformation proceeds to rise. After the experiment, the spring returns to its original unstrained form, thus creating the drop in the strain signal. Note that this kind of behavior, however, is not creep in a sense as we have defined it in Section 5.2.1.2, but instead, just flow. Hence, the Maxwell element is good at modeling stress relaxation in a viscoelastic fluid, but not creep of a viscoelastic solid.

For the latter experimental situation, as a complement to the time-dependent elastic modulus in the preceding case, a time-dependent creep compliance can be calculated:

$$\varepsilon(t) = \varepsilon_0 + \frac{\sigma_0}{\eta_0} t = \sigma_0 \left(\frac{1}{E_0} + \frac{t}{\eta_0}\right) \tag{5.30a}$$

$$J(t) = J_0 + \frac{t}{\eta_0} \tag{5.30b}$$

In a dynamic experiment, we apply a sinusoidally modulated strain to the Maxwell element. As we have learned from Section 5.3, the involved complex strain, $\varepsilon^\star$, and the complex stress resulting from it, $\sigma^\star$, can be expressed as exponential functions, of which we also need the derivatives in the following:

$$\varepsilon^\star = \varepsilon_0 \exp(i\omega t) \Rightarrow \frac{\mathrm{d}\varepsilon^\star}{\mathrm{d}t} = i\omega\varepsilon_0 \exp(i\omega t) \tag{5.31a}$$

$$\sigma^\star = \sigma_0 \exp(i(\omega t + \delta)) \Rightarrow \frac{\mathrm{d}\sigma^\star}{\mathrm{d}t} = i\omega\sigma_0 \exp(i(\omega t + \delta)) \tag{5.31b}$$

The starting point for mathematical modeling of the Maxwell element under the condition of a dynamic experiment is again eq. (5.27), which we expand by a factor of $1 = \eta/\eta$ on the right side, as this allows us to rearrange the equation as follows:

$$\frac{\mathrm{d}\varepsilon}{\mathrm{d}t} = \frac{1}{E}\frac{\eta}{\eta}\frac{\mathrm{d}\sigma}{\mathrm{d}t} + \frac{\sigma}{\eta} = \frac{1}{\eta}\left(\frac{\eta}{E}\frac{\mathrm{d}\sigma}{\mathrm{d}t} + \sigma\right) \Leftrightarrow \eta\frac{\mathrm{d}\varepsilon}{\mathrm{d}t} = \frac{\eta}{E}\frac{\mathrm{d}\sigma}{\mathrm{d}t} + \sigma \tag{5.32}$$

When we apply the fundamental equation $\eta = E_0\tau_0$ to the left-hand side of eq. (5.32) along with applying a rearranged form of it, $\tau_0 = \eta/E_0$, to the right-hand side of eq. (5.32), we get

$$E_0\tau_0\frac{\mathrm{d}\varepsilon}{\mathrm{d}t} = \tau_0\frac{\mathrm{d}\sigma}{\mathrm{d}t} + \sigma \tag{5.33}$$

Plugging in the exponential functions given in eq. (5.31a) and (5.31b) yields

$$\begin{aligned} E_0\tau_0 i\omega\varepsilon_0 \exp(i\omega t) &= \tau_0 i\omega\sigma_0 \exp(i(\omega t + \delta)) + \sigma_0 \exp(i(\omega t + \delta)) \\ &= \sigma_0 \exp(i(\omega t + \delta))(\tau_0 i\omega + 1) \end{aligned} \tag{5.34}$$

That can be rearranged as follows:

$$\frac{E_0\tau_0 i\omega}{\tau_0 i\omega + 1} = \frac{\sigma_0 \exp(i(\omega t + \delta))}{\varepsilon_0 \exp(i\omega t)} = \frac{\sigma}{\varepsilon} = E^\star \tag{5.35}$$

Expansion of the left-hand side of eq. (5.35) with the conjugate complex number allows us to write out the real and imaginary parts of the right-side, $E^\star = E' + iE''$:

$$E' = \frac{E_0\tau_0^2\omega^2}{\tau_0^2\omega^2 + 1} \tag{5.36a}$$

$$E'' = \frac{E_0\tau_0\omega}{\tau_0^2\omega^2 + 1} \tag{5.36b}$$

Analogous calculations can be made for the complex oscillatory shear modulus $G^\star = G' + iG''$.

Let us examine the latter two equations more closely by first focusing on two extreme frequency regimes, a high and a low one.

At low frequencies, the denominators of both eq. (5.36a) and (5.36b) have a value close to 1, because the $\tau_0{}^2\omega^2$ term in them becomes very small then, and therefore negligible compared to the 1 in them. Then, the power-law form of the remaining numerators of eq. (5.36a) and (5.36b) gives straight-line graphs when plotted in a double-logarithmic representation, with a slope determined by the exponent of the variable ω. Figure 76 shows such a log–log plot, in which we find a slope of 2 for E', and a slope of 1 for E''.

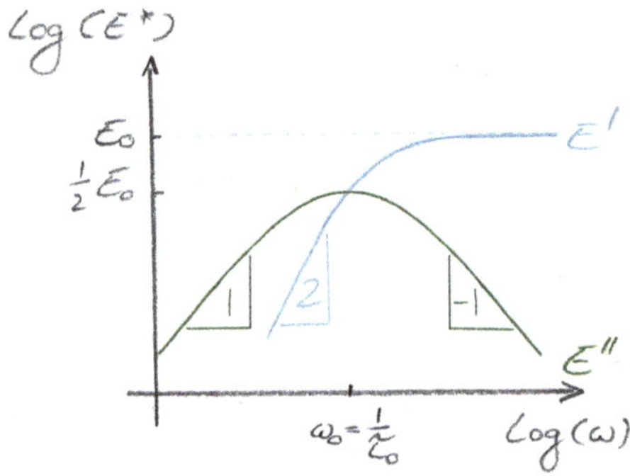

Figure 76: Frequency-dependent elastic modulus of a Maxwell element, $E(\omega)$, in a double-logarithmic representation. The graphs can be separated into two frequency regimes, separated by the crossover point of E' and E''. At low ω, the material has enough time to relax, and its viscoelastic properties are dominated by its viscous contributions, so $E'' > E'$. At high ω, the material cannot fully relax anymore, and its viscoelastic properties are dominated by its elastic contribution, so $E'' < E'$.

At high frequencies, by contrast, the $\tau_0{}^2\omega^2$ term in the denominators of both eq. (5.36a) and (5.36b) is much larger than the 1 in them, such that the latter can be neglected. Many of the variables in the numerators and in the denominators cancel out then. For E', no dependence of ω remains due to that cancelling, such that its graph is a flat line with a value of E_0 at high frequencies. For E'', we obtain a frequency dependence of ω^{-1} due to that cancelling, such that its graph is a straight line with slope −1 in a double-logarithmic representation such as the one shown in Figure 76.

Both frequency regimes are separated by the **terminal relaxation time**, τ_0, at a frequency of $\omega_0 = 1/\tau_0$. At that frequency, according to eq. (5.36a) and (5.36b), both E' and E'' have values of $\frac{1}{2}E_0$, as also seen in Figure 76; hence, both the viscous and elastic part contribute equally here.

Now let us do what we have done several times already in the course of this book and use this mathematical insight to obtain conceptual insight into the time-dependent mechanical properties of a viscoelastic Maxwell-type fluid. Again, we discuss this in terms of a low- and a high-frequency regime.

At low frequencies, there is much time for relaxation of the system. A deformed material then has enough time to rearrange its building blocks on a microscopic scale; for example, if it is a polymer system, the chains have enough time to relax back to their equilibrium conformations after deformation. Reconsidering the Maxwell element from Figure 74, the time is long enough for the damper to relax the energy stored in the spring. As a result, at such long timescales, the material's mechanical properties are dominated by its viscous contribution. This becomes more and more dominant the longer the timescale is or, in turn, the lower the frequency is. We see that clearly in Figure 76: the lower the frequency, the more does the E''-curve outweigh the E'-curve. At high frequencies, the picture is to the contrary: here, there is not enough time for relaxation of the system. Visually speaking, the damper does not have time to be active. On these short timescales, the material therefore behaves as an elastic solid. This becomes more and more dominant the shorter the timescale is or, in turn, the higher the frequency is. Again, we see that clearly in Figure 76: the higher the frequency, the more does the E'-curve outweigh the E''-curve. The transition from the low-frequency viscous regime to the high-frequency elastic regime happens at the terminal relaxation time, τ_0. It denotes the time it takes for the building blocks of any material, which in our case are polymer chains, to rearrange themselves over a distance equal to their own size. On timescales longer than τ_0, such effective displacement of the microscopic building blocks against one another is possible, causing flow on the macroscale. On timescales shorter than τ_0, such displacement is not possible, and the material responds elastically. When the time is exactly τ_0, then both the viscous and the elastic parts of the sample's mechanical spectrum contribute equally, and E' and E'' have identical values of $E_0/2$. We can observe this as the crossover point of the graphs in Figure 76.

Polymers and colloids, which are composed of large building blocks with high molar masses, exhibit τ_0 in the milliseconds to seconds range. This is advantageous, because it makes these relaxation processes observable. The same basic processes are also present in common low molar-mass materials, such as water, but in those materials, τ_0 is in the nanosecond range. These timescales are much harder or even impossible to assess experimentally. Fortunately, however, the knowledge gained from observation of the slower polymeric or colloidal systems can be adapted to the hard-to-observe classical materials due to their common physical grounds. This insight of the French physicist Pierre Gilles de Gennes was awarded with the Nobel Prize in Physics in 1991. In addition to leading to that scientific breakthrough, the millisecond-to-second timescale of relaxation of polymeric and colloidal matter makes these rich in mechanical behavior on practically relevant timescales, opening a plethora of mechanical application possibilities. It is for these that polymers are in fact most famous.

5.6.2 Viscoelastic solids: the Kelvin–Voigt model

The Maxwell model is not the only possible combination of a Hookian spring and a Newtonian dashpot. When the two are connected in parallel, we generate a system that is the basis of the **Kelvin–Voigt model**, as depicted in Figure 77.[55] Upon application of a constant stress, the Kelvin–Voigt element exhibits creep and deforms at a decreasing rate, asymptotically approaching the steady-state strain σ_0/E. When the stress is released, the material gradually relaxes to its undeformed state. A graphical representation of this experiment is shown in Figure 78. Since the deformation is reversible (though not suddenly), the Kelvin–Voigt model describes an elastic solid that also exhibits some viscous contribution.

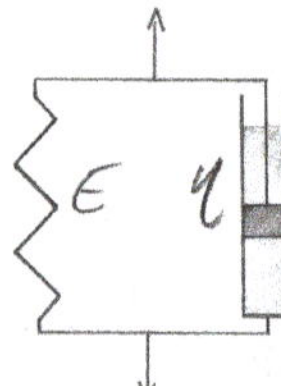

Figure 77: Kelvin–Voight element composed of a purely viscous damper and a purely elastic spring connected in series.

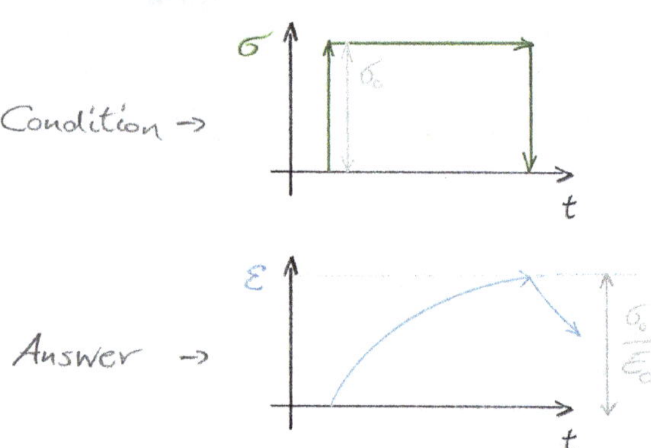

Figure 78: Creep test of a *Kelvin–Voigt element.* The response of the spring is delayed by the simultaneous but slow response of the damper. The final deformation will eventually approach that for the pure elastic part, σ_0/E. Once the stress is released, the entire Kelvin–Voigt element slowly relaxes to its original position, which means that the deformation is reversible.

55 The model was originally proposed by Oskar Emil Meyer in 1874 in his essay "Zur Theorie der inneren Reibung", published in the *Journal for Pure and Applied Mathematics.* It was later independently "rediscovered" by the British physicist William Thomson, who later became the first Baron Kelvin and by the German physicist Woldemar Voight in 1892.

By simple geometrical need, the total strain of the Kelvin–Voigt element is the same as the individual strain of each element: $\varepsilon = \varepsilon_1 = \varepsilon_2$, whereas the total stress is the sum of the individual stresses: $\sigma = \sigma_1 + \sigma_2$.

We can express the elastic contribution to the Kelvin–Voigt body by Hooke's law as

$$\sigma_1 = E\varepsilon \tag{5.37}$$

and the viscous contribution by Newton's law as

$$\sigma_2 = \eta \frac{\mathrm{d}\varepsilon}{\mathrm{d}t} \tag{5.38}$$

We can now plug the latter two equations into $\sigma = \sigma_1 + \sigma_2$ to generate

$$\frac{\mathrm{d}\varepsilon}{\mathrm{d}t} = \frac{\sigma}{\eta} - \frac{E\varepsilon}{\eta} \tag{5.39}$$

Solving this differential equation with the boundary condition of a constant applied stress of $\sigma = \sigma_0$ yields

$$\varepsilon(t) = \frac{\sigma_0}{E}\left(1 - \exp\left(-\frac{Et}{\eta}\right)\right) = \frac{\sigma_0}{E}(1 - \exp(-\tau t)) \tag{5.40}$$

Here, τ is the **retardation time**: it quantifies the delayed response to an applied stress and is a material-specific parameter.

The Kelvin–Voigt model is good at predicting creep, because in the infinite time limit the strain approaches a constant value of $\lim_{t\to\infty} \varepsilon = \frac{\sigma_0}{E}$, whereas the Maxwell model predicts an infinite linear relationship between strain and time. Thus, the Kelvin–Voigt model element is good to model a viscoelastic solid. By contrast, as we have seen above, the Maxwell model is good to model a viscoelastic liquid.

5.6.3 More complex approaches

By addition of further Hookian spring and Newtonian damper elements, a plethora of more complex models have been created to describe the viscoelastic properties of many substances. These models include but are not limited to the *standard linear solid* or *Zehner model*, the *generalized Maxwell model*, the *Lethersich*, the *Jeffreys*, and the *Burgers model*. The latter one, which is depicted in Figure 79, combines a Kelvin–Voigt with a Maxwell element by connecting them in series. It was developed by the Dutch physicist Johannes Martinus Burgers in 1935 to model the behavior of bitumen. It is sometimes also applied to polymer-based materials.

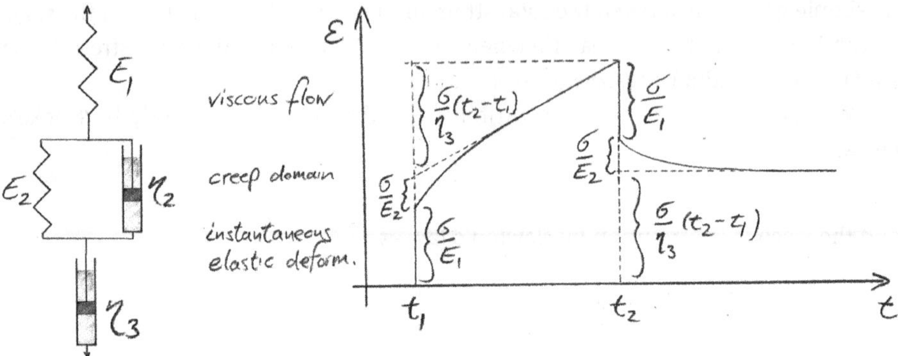

Figure 79: The Burgers model of viscoelastic media by a combination of a Kelvin–Voigt and a Maxwell element. It is tailored to accurately predict the viscoelastic properties of bitumen, but is sometimes applied to polymers as well.

5.7 Superposition principles

Lesson 13: Superposition Principles

Παντα ρηει – everything flows. This fundamental aphorism of rheology holds for polymers in particular, because even if their momentary appearance may be that of an elastic solid, in many cases, waiting long enough will let them flow. In this lesson, you will see which fundamental physical principle lies behind this, and how it allows us to superimpose data collected in different experiments to generate mechanical spectra spanning multiple decades in time or frequency.

Up to this point, we have looked upon some simplified models to describe the mechanical response of a viscoelastic material. But what do we observe in a real experiment? Let us look at a typical amorphous polymer and consider its creep compliance, J, as a function of time, t, for a range of temperatures, T, in Figure 80. (Note that J and t are the inverse quantities of E and ω, respectively.) At high temperatures, J has high values and increases proportionally with time t, which means that the material exhibits purely viscous flow. At low temperatures, by contrast, J has low values and is independent of t, which means that the material shows a purely elastic response. At intermediate temperatures, J has intermediate values, and it is first independent of

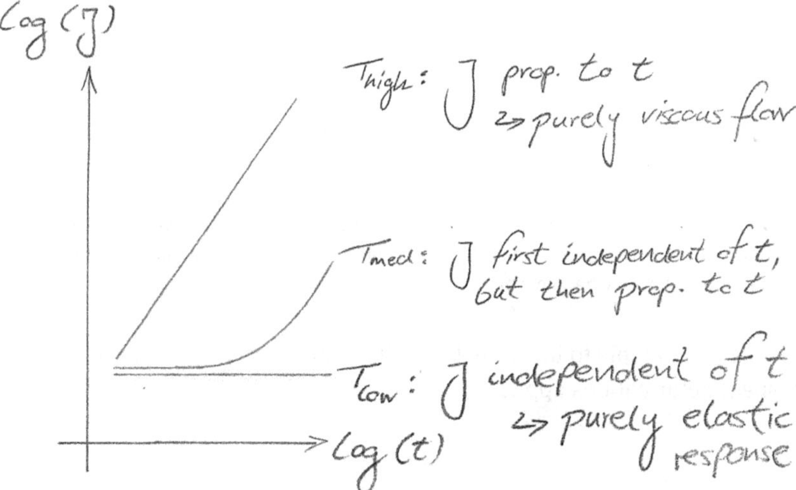

Figure 80: Creep compliance, J, of an amorphous polymer probed as a function of time, t, at various temperatures. At high temperature, J is proportional to t, meaning that the material exhibits purely viscous flow. At low temperature, by contrast, J is fully independent of t, which means that the material shows a purely elastic response. At intermediate temperatures, J is first independent of t and then becomes proportional to it.

t but then becomes proportional to it. To sum up, we see that the polymer displays a rich mechanical behavior depending on both time, *t*, and temperature, *T*.

In this section, we will see how the quantities *t* and *T* are related to each other in a dataset such as the one shown in Figure 80, and, while doing so, we will also learn some fundamental principles of the overlay, or superposition, of rheological data.

5.7.1 The Boltzmann superposition principle

The **Boltzmann superposition principle** formulates that the current state of stress or deformation of a viscoelastic material is the result of its history.[56] In mathematical terms, this means that the total stress or deformation applied to a material is the sum of the time-dependent evolution of every single stress or deformation increment that it has experienced. In other words, the system remembers any former stresses or deformations and continues to relax or creep from them even when new ones are applied on top, as depicted in Figure 81.

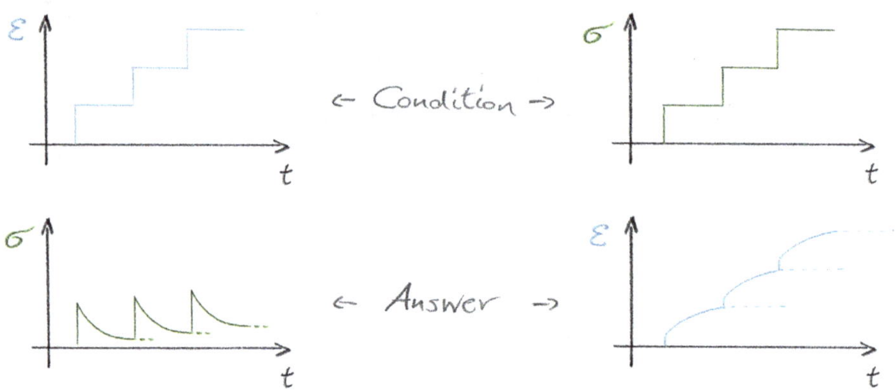

Figure 81: When multiple relaxation experiments (left) or creep tests (right) are performed in a sequence, the system remembers the stresses or deformations of each single one and continues to creep or relax from these even when new ones are applied on top.

Let us consider an example to illustrate the Boltzmann Superposition principle. The sum of all shear deformations applied to a material can be expressed as

$$\gamma = \sum_i \gamma_i \tag{5.41}$$

56 In a philosophical view, this is just like with us as individuals: our histories have made us, or "shaped us", into the persons we are today.

According to Hooke's law, the resulting sum of stresses can be expressed by the sum of the shear modulus G multiplied by the individual deformation γ_i:

$$\sigma = \sum_i \sigma_i = \sum_i G(t - t_i)\gamma_i \tag{5.42}$$

Here, t is the time of measurement, whereas t_i is the time of application of the strain increment γ_i.

We can derive an equation for the case of continuous change by integration:

$$\sigma(t) = \int \mathrm{d}\sigma = \int_{-\infty}^{t} G(t - t')\mathrm{d}\gamma = \int_{-\infty}^{t} G(t - t')\frac{\mathrm{d}\gamma}{\mathrm{d}t}\mathrm{d}t' \tag{5.43a}$$

To simplify this expression, we use the method of substitution and write $t - t' \equiv u$ to generate

$$\sigma(t) = -\int_{-\infty}^{0} G(u)\dot{\gamma}\mathrm{d}u = +\int_{0}^{\infty} G(u)\dot{\gamma}\,\mathrm{d}u \tag{5.43b}$$

This can be solved if we have an expression of the time-dependent modulus at hand. As a simple example, we take the relaxation function $G(u) = G_0 \exp(-\frac{u}{\tau})$ from the Maxwell model and insert it:

$$\sigma(t) = G_0 \int_{0}^{\infty} \exp\left(-\frac{u}{\tau}\right)\dot{\gamma}\,\mathrm{d}u \tag{5.44}$$

In a continuous-shear scenario, the shear rate $\dot{\gamma}$ is constant. For a monodisperse polymer sample, this yields

$$\sigma(t) = G_0\dot{\gamma}\int_{0}^{\infty} \exp\left(-\frac{u}{\tau}\right)\mathrm{d}u = G_0\dot{\gamma}\left[-\tau\exp\left(-\frac{u}{\tau}\right)\right]_0^{\infty} = G_0\dot{\gamma}\tau \tag{5.45}$$

We can rearrange that equation to derive an expression for the stress divided by the shear rate, which, according to Newton's law, is the viscosity η:

$$\frac{\sigma}{\dot{\gamma}} \equiv \eta = G_0\tau \tag{5.46}$$

Equation (5.46) is a very fundamental relation in the field of rheology, because it connects the macroscopic properties viscosity, η, and shear modulus, G, to the microscopic property of the relaxation time, τ, that contains information about the structure of the system's building blocks (here: polymer coils).[57] With that, we have a structure–property relation. This relation defines flow as a relaxation process, in

57 Here at this point, we have derived the fundamental identity $\eta = G\,\tau$ "properly", whereas before, in Section 5.6.1, it was obtained more as a "mathematical by-product".

which the viscosity is composed of the relaxation modulus and the relaxation time. We can understand that quite illustratively: viscosity (η) quantifies the ability of a fluid medium to initially store energy upon shear deformation (G) times how long it takes to dissipate that by subsequent relaxation (τ), that means, by positional change of the fluid's building blocks (molecules or particles). The higher both these contributors are, the harder it is to stir or pump the fluid.

In the more complex case of a polydisperse relaxation time spectrum, the viscosity is expressed as the sum of multiple relaxation contributions:

$$G(u) = \sum_i G_i \exp\left(-\frac{u}{\tau_i}\right) \Rightarrow \frac{\sigma}{\dot{\gamma}} \equiv \eta = \sum_i G_i \tau_i \tag{5.47}$$

This is the case when either the system's building blocks are polydisperse in size, or if monodisperse building blocks exhibit *multiple*, hierarchical relaxations, such as coil-internal Rouse modes.

In summary, we see that relaxation occurs on the timescale τ, or on a time spectrum of multiple overlaying τ_i, and that these determine the viscosity by the elementary formula $\eta = G_0\tau$. In the next section, we will take a closer look at the temperature dependence of that.

5.7.2 The thermal activation of relaxation processes

Flow requires the molecules or particles that constitute a material to change their positions. This necessitates them to slip by each other, for which they have to overcome an energetic activation; it also necessitates some free volume around them to move into. Both these prerequisites are temperature dependent, which means that, conceptually, they can be treated together as an effective activation energy barrier, $E_{a,eff}$, that determines the temperature-dependent frequency with which positional changes of the molecules occur, $1/\tau$, in the form of an Arrhenius-type expression:[58]

$$\frac{1}{\tau} \sim \eta \sim \exp\left(\frac{E_a}{k_B T}\right) \cdot \exp\left(\frac{V^*}{V_f}\right) \approx \exp\left(\frac{E_{a,\mathrm{eff}}}{k_B T}\right) \tag{5.48}$$

In eq. (5.48), the first term features the activation energy, E_a, for each molecular or particulate positional change past other molecules or particles, whereas in the second term, V^*/V_f reflects the probability to find enough free space in the vicinity during

58 This is analogous to the rate constant k of a chemical reaction, which also increases with temperature according to an Arrhenius interrelation, because the activation energy is overcome more frequently at higher temperature. The rate constant, in turn, is related to the reaction half-life time, τ, in the form of $k \sim \frac{1}{\tau}$, whereby the constant of proportionality depends on the order of the reaction.

such an attempt of positional change, assessed by the free volume in the system, V_f, and the volume required for each molecular or particulate rearrangement, $V^\star$.[59]

The effective activation energy $E_{a,eff}$ is more often overcome at high temperatures, which means that high temperatures facilitate positional change of the molecules. Thus, high temperatures lead to short relaxation times, whereas low temperatures lead to long relaxation times. The lowest possible temperature at which movement is still possible is the glass transition temperature, T_g. Below this specific temperature, thermal activation is not possible anymore, all movement is frozen, and the relaxation time is infinite, $\tau \rightarrow \infty$.

5.7.3 Time–temperature superposition

The fundamental connection of temperature to the rate of chain relaxation, which – in turn – is connected to the viscosity according to eq. (5.46), allows us to shift and superimpose viscoelastic data that were recorded at different temperatures so as all superimpose on a common time axis; this is done by shifting these data along the time axis by application of a **shift factor,** a_T:

$$G(t, T) = b_T G\left(\frac{t}{a_T}, T_0\right) \tag{5.49}$$

The left-hand side of eq. (5.49) features $G(t,T)$, which is the shear modulus measured at a time t and temperature T. The right-hand side of the equation tells us that the same modulus will also be measured at a different temperature, T_0, if the timescale is shifted by a factor a_T. Often, a suitable standard reference temperature is taken for T_0, for example, room temperature or the polymer's glass transition temperature, T_g. To also account for potential density differences of the material at the temperatures T and T_0 (due to thermal expansion), a further shift factor $b_T = \rho T / \rho_0 T_0$ is often applied as well, as also seen in eq. (5.49).

The shift factors are not just loose arbitrary quantities, but they can be conceptualized and understood to be part of further interrelations. One such relation is the **Williams–Landel–Ferry** (WLF) equation. It was originally discovered empirically, but it can also be derived from eq. (5.48) as follows: if two viscosities η_1 and η_2 are measured at two temperatures T_1 and T_2, their ratio according to eq. (5.48) is

$$\frac{\eta_1}{\eta_2} = \exp\left(\frac{V^\star}{V_{f_1}} - \frac{V^\star}{V_{f_2}}\right) \tag{5.50}$$

59 Both these contributions have first been introduced independent of one another according to the free volume theory by Doolittle (*J. Appl. Phys.* **1951**, *22*(12), 1471–1475) and the theory of rate processes by Eyring; later on, however, it was recognized that they must both be accounted together by Macedo and Litovitz (*J. Chem. Phys.* **1965**, *42*(1), 245–256).

if we assume that the temperatures T_1 and T_2 are high enough such that the influence of the temperature on the fundamental positional change process can be neglected ($\frac{E_a}{k_B T_1} - \frac{E_a}{k_B T_2} \approx 0$). Furthermore, the relationship between the free volumes at T_1 and T_2 is as follows:

$$V_{f_2} = V_{f_1} + \Delta\alpha V_m (T_2 - T_1) \tag{5.51}$$

Here, $\Delta\alpha$ is the difference between the material's thermal expansion coefficients at both temperatures, and V_m is the eigenvolume of the chain material.

Logarithmization of eq. (5.50) as well as rearrangement and insertion of eq. (5.51) yields

$$\ln \frac{\eta_1}{\eta_2} = \frac{V^*}{V_{f_1}} \cdot \frac{T_2 - T_1}{\left(\frac{V_{f_2}}{\Delta\alpha V_m}\right) + (T_2 - T_1)} \tag{5.52}$$

If T_1 is replaced by the glass transition temperature of the material under consideration, T_g, the equation for any second temperature $T = T_2$ is

$$\ln \frac{\eta_g}{\eta_T} = \frac{\left(\frac{V^*}{V_{f,g}}\right) \cdot (T - T_g)}{\left(\frac{V_{f,T}}{\Delta\alpha V_m}\right) + (T - T_g)} \tag{5.53}$$

Based on temperature-dependent viscosity measurements, Williams, Landel, and Ferry estimated the volume V^* required for a microscopic positional change as $V^* \approx 40 \cdot V_{f,g}$ and the free volume at the glass transition temperature, $V_{f,g}$, as $V_{f,g} \approx 52 \cdot \Delta\alpha V_m$. Furthermore, in a polymer melt, V^* can be approximately equated with the eigenvolume of the chain material V_m. If we do that, then the stepwise change in the thermal expansion coefficient $\Delta\alpha$ is determined to be $\Delta\alpha = 4.8 \cdot 10^{-4}\ \mathrm{K}^{-1}$.

By plugging in these values into eq. (5.53), we obtain the WLF equation that relates the shift factor a_T at temperature T to the glass-transition reference state:

$$\log a_T = \log \frac{\tau}{\tau_g} \cong \log \frac{\eta(T)}{\eta(T_g)} = \frac{-c_1 (T - T_g)}{c_2 + (T - T_g)} = \frac{-17.44 (T - T_g)}{51.6\mathrm{K} + (T - T_g)} \tag{5.54}$$

In the latter form, we have further transferred to decimal logarithms ($\log x = \ln x \cdot \log e$).

The numbers $c_1 = 17.44$ and $c_2 = 51.6$ K in eq. (5.54) are universal for many different polymers if T_g is taken as the reference temperature.

The utility of the shift factors introduced in eq. (5.49), which we may either determine empirically or calculate with the WLF eq. (5.54), is that they allow us to create a master curve of many individual datasets of $G(T, t)$ referenced to new variables $G(T_0, t/a_T)$, as shown in Figure 82 (note that this plot has a frequency axis instead of a time axis, which is just the inverse). Such a master curve may then display the full rheological spectrum of a polymer, covering many orders of magnitude in G (from single pascals to gigapascals!) and t (or ω) (from nano- or milliseconds to decades!), which would be inaccessible by actual experimentation.

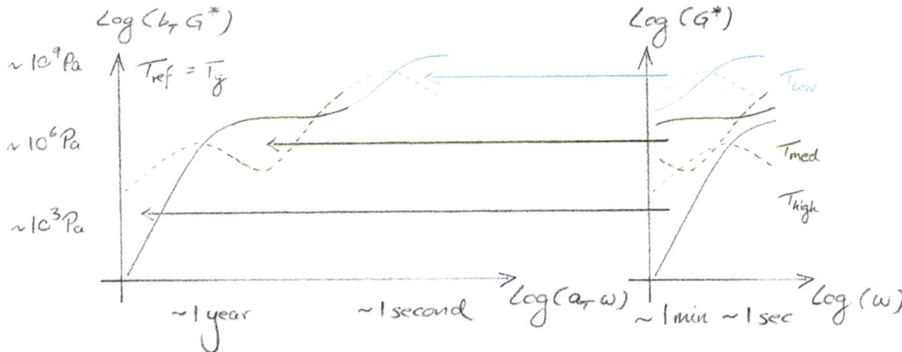

Figure 82: Construction of a rheological master curve according to eq. (5.49). Full lines denote the storage modulus, G', whereas dashed lines denote the loss modulus, G''. The reference temperature in this example is the glass transition temperature. By assembling the single curves obtained at various temperatures over a frequency span of about three decades, as collected in the right part of the figure, and shifting them along the frequency axis through application of suitable shift factors a_T, along with potential additional shifting along the modulus axis to account for density differences at different temperatures through the use of shift factors b_T, we can create a master curve spanning 12(!) orders of magnitude on the time axis, as shown in the left part of the figure.

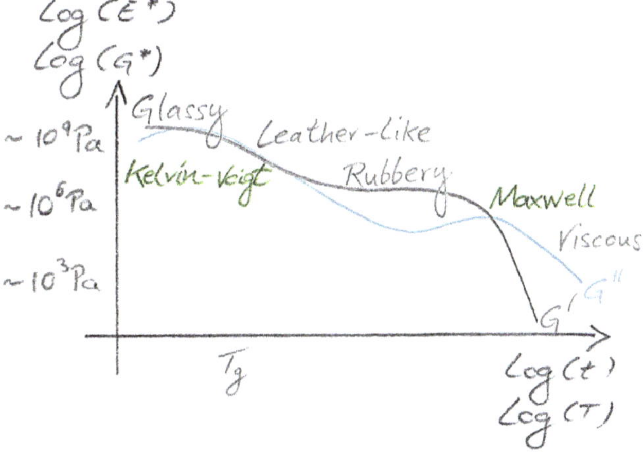

Figure 83: Schematic representation of a polymer's full mechanical spectrum. At low temperatures or on very short times, the polymer is a glassy, energy elastic solid. When the temperature is raised or the time extended, the polymer enters a leather-like regime in which it is a viscoelastic solid described by the Kelvin–Voigt model. For samples that are composed of long chains, we then observe an intermediate rubbery plateau regime. When the temperature is sufficiently high or the time sufficiently long, the material will eventually flow like a viscoelastic liquid described by the Maxwell model. The Pascal values on the ordinate are typical for polymers in the melt state.

In such a full rheological spectrum, we can recognize several distinctly different regimes, depending on the timescale and on the temperature of observation. Figures 83 and 84 display these different regimes in a schematic fashion. Note that Figure 83 is flipped compared to both Figures 82 and 84, because it shows the moduli E and G as a function of time t rather than frequency ω. However, this is just transformation of the type of displaying, but the characteristic domains remain identical in either representation.

5.8 Viscoelastic states of a polymer system

Lesson 14: Mechanical Spectra

The principle of time–temperature superposition that you got to know in the last lesson allows master curves of the time- or frequency-dependent moduli of polymer samples to be composed. Such rheological spectra feature multiple different domains, depending on in which state of mobility or trapping the chains in the sample are. The following lesson will give an overview of these states and summarize the rich variety of viscoelastic states of polymers.

5.8.1 Qualitative discussion of the mechanical spectra

Let us discuss the schematic of a polymer viscoelastic spectrum in Figure 83 from left to right along the time axis (that can also be viewed as a temperature axis due to the principle of time–temperature superposition), or alternatively, the variant schematic in Figure 84 from the right to the left on the frequency axis (which is nothing else than a flipped time axis on log scales). At very short timescales, corresponding to very high frequencies and to low temperatures, the elastic modulus has high values in the range of gigapascals, and its storage part, G' or E', exceeds its loss part, G'' or E''. This means that the material is a tough elastic solid in this regime. A molecular-scale interpretation is that on these very short timescales, or at that low temperatures, no motion is possible at all in the sample, neither of the polymer chains as a whole nor of just parts of them. This is because the temperature is either so low that any motion is frozen, or because we consider timescales so short that these motions can just not be achieved yet. As a result, we have a sample with amorphous structure and without any dynamics: a **glass**. This glassy state features energy-elastic mechanics with a high modulus. At a first characteristic time, τ_0, the shortest possible relaxation time (which is typically around 1 ns at room temperature for a polymer melt), at least single monomeric units have enough time to relax in a sense that they can diffuse over distances equal to their own size. At times longer than that, more and more additional relaxation modes are activated, in a sense that now also sequences of pairs, triples, quadruples, and so on of monomers can be displaced over distances equal to their own size. This sequential activation of more and more relaxation modes more and more drops the modulus if we go from left to right on the time axis in Figure 83 or from right to left on the frequency axis in Figure 84. The polymer is then in a markedly **viscoelastic** *leather-like regime*, as we see in Figures 83 and 84, where both the storage and loss moduli, G' and G'', have very similar values and very similar time- or frequency-dependent scaling. The mechanical properties in that state can be treated analytically by the Rouse model (for modeling the polymer subchain relaxation modes and the dropping of the time- or frequency-dependent modulus as more and more of these modes are activated) or phenomenologically by the Kelvin–Voigt model (to phenomenologically treat the viscoelastic solid-like appearance of the specimen).

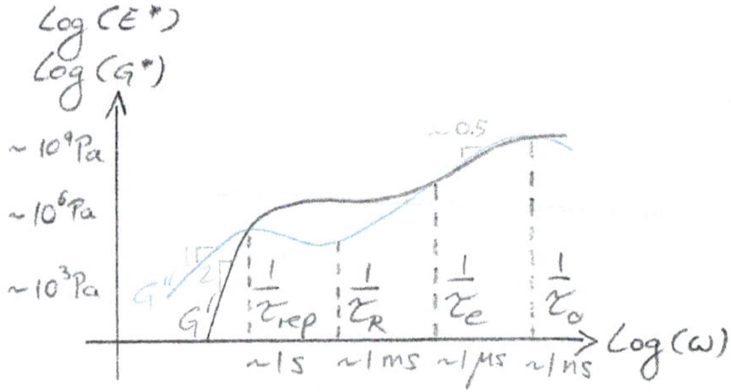

Figure 84: Full rheological spectrum of a polymer melt at room temperature. At the shortest possible relaxation time, τ_0, which lays on the right extreme of the spectrum shown in the present frequency-dependent form, only subchains of single monomeric units can relax; following upon that, at frequencies left of that extreme, comes a viscoelastic regime in which sequential activation of relaxation modes of pairs, triples, quadruples, and so on of monomer units leads to dropping of the moduli. At the entanglement time, τ_e, chain segments realize mutual entanglement, leading to an intermediate elastic plateau. From the Rouse time on, τ_R, there is sufficient time for the entire chain to be displaced, such that it could actually fully and freely relax and flow already if it would not be trapped by entanglements. Getting loose of that constraint is eventually possible on timescales longer than the reptation time, τ_{rep}. From that time on, on the very left end of the spectrum shown here, the material is a viscoelastic liquid phenomenologically described by the Maxwell model, with $G' \sim \omega^2$ and $G'' \sim \omega$.

At the end of the viscoelastic leather-like regime that features just subchain motion, intuitively, we would expect to observe a transition to a free-flowing regime in which relaxation and displacement of the whole chains can occur. From the Rouse (and also from the Zimm) model of polymer dynamics, we know that this should happen on times longer than the Rouse (or the Zimm) time, τ_R (or τ_Z), which is denoted to be about 1 ms in Figure 84. Such a transition will indeed be observed in the case of samples with short chains. In samples with long chains, by contrast, such as those shown in Figure 82–84, before that free-flow regime, we observe something else: an intermediate **rubbery elastic plateau**. This plateau starts at a second characteristic time: the entanglement time τ_e, which is around 1 μs in Figure 84. At that time, segments of long chains realize that they are trapped by mutual entanglement. This constraint hinders them from relaxing fully and freely, thereby leading to a regime in which deformation energy cannot be relaxed but is stored, with a modulus in the range of about a megapascal. As in that regime, motion of the chains is already activated but topological entanglement still hinders them to relax, deformation of the sample is accommodated by decoiling of the chains; as a result, the elasticity in this regime is of entropic origin. When the time is eventually long enough for the polymer chains to disentangle and move by a mechanism named **reptation** on timescales longer than the reptation time τ_{rep}, which is around 1 s in Figure 84, the polymer shows **terminal viscous flow**.

In this regime, the material is a *viscoelastic liquid* phenomenologically described by the Maxwell model, with G' scaling with ω^2 and G'' scaling with ω. From then on, the more time we give the material, the more will G'' exceed G', meaning that the more will the material mechanics be dominated by its viscous properties.

5.8.2 Quantitative discussion of the mechanical spectra

Now that we are able to visualize the full mechanical spectrum of a polymer and to describe it phenomenologically, let us discuss these spectra quantitatively. For this purpose, we first focus on unentangled polymer melts or solutions that both do not display any intermediate rubbery plateau. We know from Section 5.6.1 that the mechanics of viscoelastic liquids is described by the Maxwell model, and we know that the dynamics of polymer chains is quantified by the Rouse and the Zimm model. We have also just learned that, in addition to the chain as a whole on timescales longer than the Rouse or Zimm time, smaller chain segments of various lengths can relax on shorter timescales. These subchain relaxations are appraised by so-called relaxation modes numerated by an index p. The pth mode corresponds to the coherent motion of a subchain with N/p segments if N is the total number of monomer segments in the whole chain. This means that at $p = 1$, the entire chain relaxes and can get displaced by a distance equal to its own size, whereas at $p = 2$, only each half of it relaxes and can get displaced by a distance equal to the half-chain size, respectively. At $p = 3$, just only each third of the chain relaxes and can get displaced by a distance equal to one-third of the whole chain size, and so forth. At $p = N$, only single monomeric units can relax and get displaced against each other by their own size. Figure 85 visualizes this hierarchy of relaxation modes. At a time τ_p after abrupt deformation, all modes with index above p are relaxed already, whereas all modes with index below p are still unrelaxed. Each of the p subchains that belong to the mode with index p relax like own independent chains of length N/p, with a relaxation time that can be appraised by the Rouse or Zimm formalism if this were applied to a chain with not N segments, as usual, but only N/p segments; this relaxation time may also be plugged into a Maxwell-type formalism to phenomenologically model the mechanical spectra of these subchains. Superposition of all these possible relaxation spectra yields

$$G'(\omega) = \nu k_\text{B} T \sum_{p=1}^{N} \frac{{\tau_p}^2 \omega^2}{{\tau_p}^2 \omega^2 + 1} \tag{5.55a}$$

$$G''(\omega) = \nu k_\text{B} T \sum_{p=1}^{N} \frac{\tau_p \omega}{{\tau_p}^2 \omega^2 + 1} \tag{5.55b}$$

Here, ν is the number-per-volume concentration of chains in the sample, whereas N is the number of monomer units per chain, and with that, also the total number of possible relaxation modes per chain.

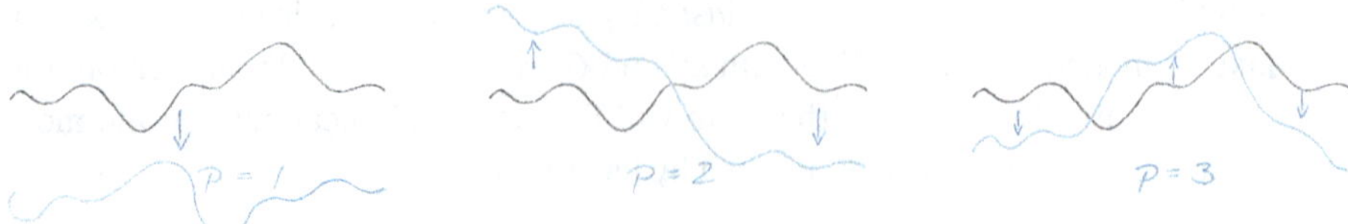

Figure 85: Relaxation modes, indexed by a number p, of a schematic polymer chain. The first mode, $p = 1$, relates to relaxation of the entire chain. In the second mode, $p = 2$, subchain segments with a length of just half of the chain can relax. The third mode, $p = 3$, corresponds to the relaxation of subchain segments with a length of only a third of the chain, and so on. In the last mode, $p = N$, only single monomeric units can relax (not sketched here). Picture modified from H. G. Elias: *Makromoleküle, Bd. 2: Physikalische Strukturen und Eigenschaften* (6. Ed.), Wiley VCH, **2001**.

We can see from this expression that at frequencies $\omega < 1/\tau_1$, all coil-internal modes can relax. G' then scales with ω^2 and G'' scales with ω, because the denominators in eq. (5.55) can be assumed to be one for such small ω-values, as the $\tau_p{}^2\omega^2$ parts in them become negligible then. At frequencies $\omega > 1/\tau_1$, by contrast, this changes: now, each unrelaxed mode contributes an energy-storage increment of k_BT to the moduli. The moduli therefore increase from k_BT per chain in the sample volume at τ_1 ($=\tau_{Rouse}$ or τ_{Zimm}) to k_BT per monomer in the sample volume at τ_N ($=\tau_0$) (note: by the normalization of these energies to the sample volume, we obtain a quantity with unit $Nm/m^3 = N/m^2 = Pa$). In that regime, the time dependence of G' and G'' (or inversely, the frequency dependence of these moduli) is therefore given by the time (or frequency) dependence of the mode index. This can be obtained from the Rouse model for polymer melts or the Zimm model for polymer solutions, as summarized in Table 8 in Section 3.6.4. According to eq. (5.55a) and (5.55b), we can calculate the frequency-dependent storage and loss moduli for the individual modes p, G_p' and G_p'', and visualize how they contribute to the overall storage and loss moduli, G' and G''. This is shown in Figure 86 for an ideal polymer with N Kuhn monomer units. We see that at times longer than τ_1 ($=\tau_{Rouse}$ or τ_{Zimm}), there is enough time for the entire chain to relax by thermal motion, such that the energy-storage capability is just k_BT per chain. At a shorter time, τ_2, subsegments with a length of up to only half of the entire chain can relax, whereas the mode τ_1 is still unrelaxed. The energy-storage contribution to the modulus is now k_BT for each unrelaxed mode. This trend continues: the shorter the time, the more modes are still unrelaxed, with each unrelaxed mode contributing one k_BT energy-storage increment to the modulus. The overlay of still-unrelaxed and already-relaxed modes leads to a scaling of G' and G'' with $\omega^{1/2}$. At times shorter than τ_N ($=\tau_0$), not even single-monomeric segmental relaxation is possible anymore, and the polymer chain motion is practically frozen. All the unrelaxed modes now each contribute an energy-storage increment of k_BT to the moduli, such that G' has achieved its maximum value of $G_\infty = \nu N\, k_BT$.

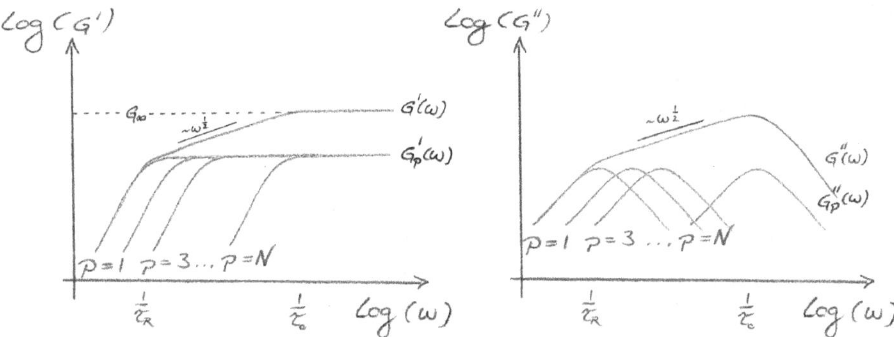

Figure 86: Frequency-dependent storage (left) and loss (right) moduli for the individual modes p, $G_p'(\omega)$ and $G_p''(\omega)$, as well as the entire chain, $G'(\omega)$ and $G''(\omega)$, of a polymer with N Kuhn segments. Redrawn from C. Wrana: *Polymerphysik*, Springer, **2014**.

Indeed, such mechanical spectra can be observed for unentangled polymer melts and solutions. The first type of sample (i.e., polymer melts) is better descried by the Rouse model, whereas the second (i.e., polymer solutions) is better described by the Zimm model. Figure 87 shows schematics of the mechanical spectra of these types of samples. Again, at times below $t = \tau_{\text{Rouse}}$ and $t = \tau_{\text{Zimm}}$, the polymer's viscoelastic properties are those of a viscoelastic liquid, with scaling according to $G'{\sim}\omega^2$ and $G''{\sim}\omega$. Here, the chains have enough time to fully relax. At intermediate times, τ_{Rouse} or $\tau_{\text{Zimm}} < t < \tau_N$, only relaxation of chain segments is possible. The material is a viscoelastic solid and both moduli, G' and G'', have similar values and both scale with ω according to power laws with exponents corresponding to the inverse time dependence (i.e., the frequency dependence) of the mode index from Table 8 in Section 3.6.4, giving values in the range of 0.5 … 0.6 if ν is set to be 0.5 (Θ-state) … 0.6 (good-solvent state). At times shorter than τ_N (=τ_0), no relaxation is possible anymore, and the material is a glassy solid.

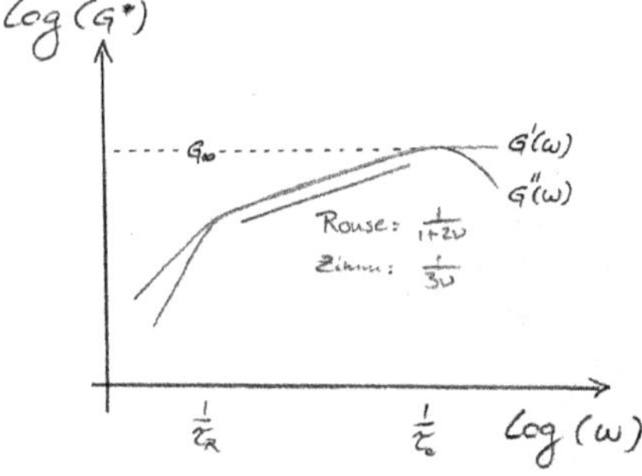

Figure 87: Schematic viscoelastic spectra of unentangled polymers.

5.9 Rubber elasticity

Lesson 15: Rubber Elasticity

In Lesson 3, you have got to know polymers as entropic springs. This principle causes polymer samples with entangled or crosslinked chains to show a characteristic plateau in the time- or frequency-dependent modulus. In this lesson, you will get to know how the height of this plateau can be conceptually and mathematically understood, and how this allows you to fundamentally connect the mesh structure of a polymer network (be it a permanent crosslinked or just a transiently entangled one) to its elastic modulus.

In the last lesson, we have examined the viscoelastic properties of short polymer chains. They transition directly from the viscoelastic leather-like regime at $\tau < \tau_{\text{Rouse}}$ to the viscous terminal-flow regime at $t > \tau_{\text{Rouse}}$, as shown in Figure 87. The picture, however, is different if the polymer chains are long enough such that they can form **entanglements** with one another and thereby mutually impair each other's relaxation. As we will see later, this is possible once the chains are longer than a certain minimal length, which is connected to a minimal degree of polymerization, N_e, or analog, a minimal molar mass, M_e, from which on chain entanglement can first occur. If chains are longer than $N_e l$ (and therewith heavier than M_e), mutual entanglement impairs their relaxation, and only on long timescales, these entangled chains can escape this mutual constraint. On short timescales, by contrast, the entanglements act like crosslinking points between the chains, thereby allowing for elastic energy storage that

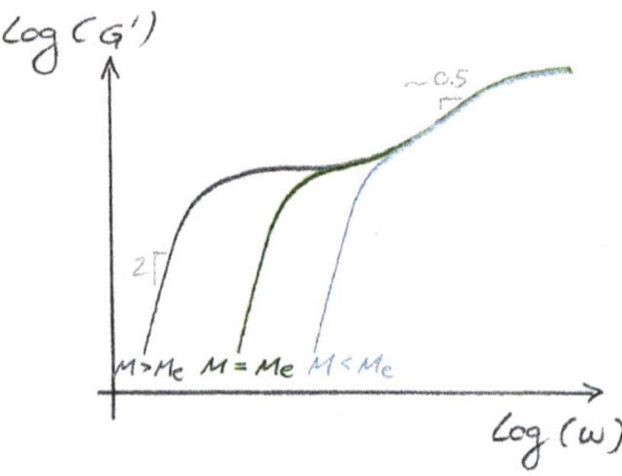

Figure 88: Mechanical spectra of chains with molar masses below (blue), at (green), and above (black) the entanglement molar mass, M_e. The latter shows a distinct rubbery plateau between the viscoelastic ($G' \sim \omega^{\frac{1}{2}}$) and the viscous ($G' \sim \omega^2$) regimes. If an even higher molar mass were shown in addition, it would display an even more extended plateau. At the high-frequency end of the spectra, all data coincide, as here, subchain relaxation modes and transition to the glassy regime is covered, which is the same for a given type of polymer at given experimental conditions, independent of the total chain length of that material.

manifests itself in the appearance of an intermediate plateau regime in the viscoelastic spectrum, the so-called **rubbery plateau** depicted in Figure 88. The longer the chains are, the longer does it take them to disentangle from one another, such that the extent of the rubbery plateau regime on the time or frequency axis is greater the higher the molar mass is. To quantitatively assess the impact of the entanglements on the mechanical properties of these long polymer chains, we regard them as permanent crosslinks and spend a focus on crosslinked polymer networks in this section.

5.9.1 Chemical thermodynamics of rubber elasticity

In chemical thermodynamics, free energy, F, is given as the relation between internal energy, U, and temperature, T, multiplied by entropy, S:

$$F = U - TS \tag{5.56}$$

The force of deformation of a material specimen f is the length derivative of this function:

$$f = \left(\frac{\partial F}{\partial l}\right) = \left(\frac{\partial U}{\partial l}\right) - T\left(\frac{\partial S}{\partial l}\right) \tag{5.57}$$

The entropy derivative in eq. (5.57) can be rewritten based on the identity $S = -(\partial F/\partial T)$ and the symmetry of second derivatives (Schwarz' theorem) for total differentials:

$$\left(\frac{\partial S}{\partial l}\right) = -\frac{\partial}{\partial l}\left(\frac{\partial F}{\partial T}\right) = -\frac{\partial}{\partial T}\left(\frac{\partial F}{\partial l}\right) = -\left(\frac{\partial f}{\partial T}\right) \tag{5.58}$$

Reinserting this into eq. (5.57) generates

$$f = \left(\frac{\partial U}{\partial l}\right) + T\left(\frac{\partial f}{\partial T}\right) \tag{5.59}$$

This expression can be plotted as a linear equation with an intercept of $\partial U/\partial l$ and a slope of $\partial f/\partial T$. Such a plot is shown for a typical rubber in Figure 89, in which the force has been normalized to the area, thereby yielding the more general quantity stress; the illustrated data can be imagined to be from an experiment in which the force to achieve a certain deformation is measured at various temperatures.

At temperatures below T_g, where all chain dynamics is frozen, the material is energy elastic. In that state, deformation pulls the motionless monomer segments apart from each other against their attractive interaction potentials. This is easier to achieve at higher temperatures, leading to less stress in the material at higher temperature upon deformation, which causes a negative slope in the left branch of the dataset in Figure 89. By contrast, above T_g, where chain dynamics is activated, the material is entropy elastic. In that state, deformation leads to uncoiling of the coiled

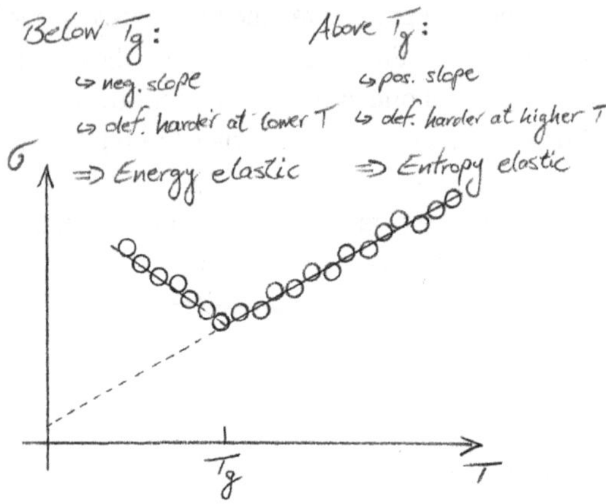

Figure 89: Relation of stress to temperature for a typical rubbery material. Below the glass transition temperature, T_g, the material is energy elastic, making it harder to deform the lower the temperature is. By contrast, above T_g, the material is entropy elastic, making it harder to deform the higher the temperature is. Picture inspired by B. Tieke: *Makromolekulare Chemie*, Wiley VCH, **1997**.

chains by transition of local *cis* into local *trans* segmental conformations. This chain uncoiling reduces the chain entropy, thereby creating an entropy-based back-driving force. Due to the fundamental coupling of temperature and entropy, that kind of deformation gets harder at higher temperature, leading to a positive slope in the right branch of the dataset in Figure 89. Conversely, the material heats up upon stretching in that state. We can see this from the following expression:

$$\left(\frac{\partial T}{\partial l}\right) = \left(\frac{\partial T}{\partial S}\right)\left(\frac{\partial S}{\partial l}\right) = \left(\frac{\partial T}{\partial H}\right)\left(\frac{\partial H}{\partial S}\right)\left(\frac{\partial S}{\partial l}\right) = -\frac{1}{c_l}T\left(\frac{\partial S}{\partial l}\right) \tag{5.60}$$

When the entropy term $\partial S/\partial l$ is negative, which it is upon deformation, the temperature term $\partial T/\partial l$ is positive, meaning that the temperature increases upon deformation.

We also see in Figure 89 that the extrapolated intercept of the positive-slope part of the dataset is small; this underlines once more that the energetic contribution to stress in a deformed rubbery sample is small.

5.9.2 Statistical thermodynamics of rubber elasticity

In addition to the chemical-thermodynamic argument above, a statistical-thermodynamic approach can give us particularly valuable quantitative information about the rubbery elastic state. In such an approach, we generally appraise the most central quantity in thermodynamics, entropy, as the degree to which a specific macrostate of interest, in our case the shape of the polymer coil, is realized by different possible microstates, in our case the local segmental *cis* or *trans* conformations. We have learned in Chapter 2 that the ideal shape of a single polymer chain is that of a Gaussian random coil, with an entropy given by random-walk statistics:

$$S = S_0 - \frac{3k_B r^2}{2\langle r^2 \rangle_0} = S_0 - \frac{3k_B r^2}{Nl^2} \tag{5.61}$$

In eq. (5.61), $\langle r^2 \rangle_0$ is the equilibrium mean-square end-to-end distance of the polymer chain, whereas r is the end-to-end distance in a given situation of interest. When we stretch the chain, we extend this latter distance, such that $r^2 > \langle r^2 \rangle_0$. Because the variable r^2 is in the numerator, whereas the constant $\langle r^2 \rangle_0$ is in the denominator of a term with negative sign in front of it, we reduce the entropy S upon increase of r, that is, upon stretching. Note that the latter equation was actually derived for single uncrosslinked polymer chains; we now presume that it also applies to each **network chain** in a **crosslinked rubber.**

Figure 90 displays the relevant parameters that are needed to treat the deformation of a polymer network sample, both on the macro- and the microscale. Due to the self-similar nature of polymers, we can express the entropy change ΔS of a single ideal network chain, which is a subchain ranging from one crosslink to another in the network, upon deformation as follows:[60]

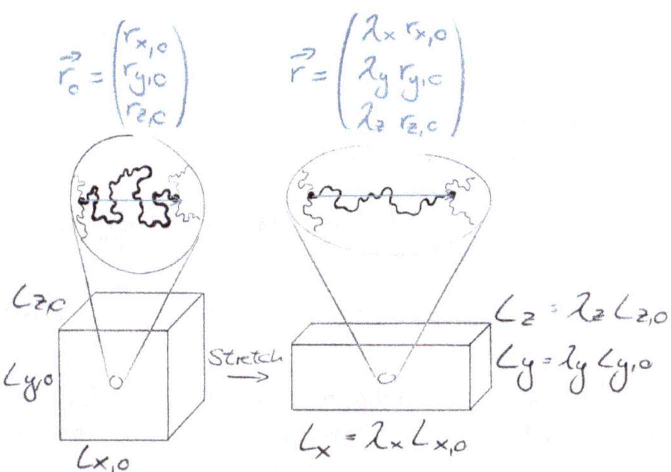

Figure 90: When a polymer-network specimen is deformed, its new dimensions L_j can be described as the original ones $L_{j,0}$ multiplied by strain ratios λ_j (with j denoting the geometrical dimension x, y, and z). In an affine scenario, we presume the microscopic deformation of the chains that constitute the sample to be equal to the macroscopic deformation of the entire body. As a result, the strain ratios λ_j are identical on both microscopic and macroscopic scales. Picture redrawn from M. Rubinstein, R. H. Colby: *Polymer Physics*, Oxford University Press, **2003.**

60 Note again that in this context, deformation means the decoiling of the polymer chain, but not stretching of the bonds between its monomeric units.

$$\Delta S = S\left(\vec{R}\right) - S\left(\vec{R}_0\right) = -\frac{3k_B\left(r_x^2 + r_y^2 + r_z^2\right)}{2\langle r^2\rangle_0} + \frac{3k_B\left(r_{x,0}^2 + r_{y,0}^2 + r_{z,0}^2\right)}{2\langle r^2\rangle_0} =$$

$$= \frac{3k_B}{2\langle r^2\rangle_0}\left(\left(\lambda_x^2 - 1\right)r_{x,0}^2 + \left(\lambda_y^2 - 1\right)r_{y,0}^2 + \left(\lambda_z^2 - 1\right)r_{z,0}^2\right) \quad (5.62)$$

The entropy change of the entire network is obtained simply by summation over all its n network chains.

$$\Delta S = -\frac{3k_B}{2\langle r^2\rangle_0}\left(\left(\lambda_x^2 - 1\right)\sum_{i=1}^{n}(r_{x,0})_i^2 + \left(\lambda_y^2 - 1\right)\sum_{i=1}^{n}(r_{y,0})_i^2 + \left(\lambda_z^2 - 1\right)\sum_{i=1}^{n}(r_{z,0})_i^2\right) \quad (5.63)$$

In an isotropic sample, the average end-to-end distances are equal in all directions:

$$\frac{1}{n}\sum_{i=1}^{n}(r_{x,0})_i^2 = \langle r_{x,0}^2\rangle = \langle r_{y,0}^2\rangle = \langle r_{z,0}^2\rangle = \frac{\langle r^2\rangle_u}{3} \quad (5.64)$$

Here, $\langle r^2\rangle_u$ is the mean-square end-to-end distance of the network chains in the *undeformed* state.

We can simply rearrange this to

$$\sum_{i=1}^{n}(r_{x,0})_i^2 = \sum_{i=1}^{n}(r_{y,0})_i^2 = \sum_{i=1}^{n}(r_{z,0})_i^2 = \frac{n\langle r^2\rangle_u}{3} \quad (5.65)$$

in which n is the number of network chains.

We can insert this expression into eq. (5.63) to generate

$$\Delta S = -\frac{3k_B}{2\langle r^2\rangle_0}\left(\left(\lambda_x^2 - 1\right)\left(\frac{n}{3}\right)\langle r^2\rangle_u + \left(\lambda_y^2 - 1\right)\left(\frac{n}{3}\right)\langle r^2\rangle_u + \left(\lambda_z^2 - 1\right)\left(\frac{n}{3}\right)\langle r^2\rangle_u\right) =$$

$$= -\frac{nk_B}{2}\frac{\langle r^2\rangle_u}{\langle r^2\rangle_0}\left(\lambda_x^2 + \lambda_y^2 + \lambda_z^2 - 3\right) \quad (5.66)$$

From that equation, it becomes clear that the entropy reduction upon stretching actually stems from two contributions. First, the entropy is reduced more if we stretch to a greater extent, which is expressed in the form of higher strain ratios λ_j in eq. (5.66). Second, the number of chains n that are stretched by that extent proportions the entropy reduction. Remember again in this context that we treat network chains, meaning chain segments ranging from one crosslinking junction to another. Consequently, a high crosslinking degree also amplifies the change of entropy, because it corresponds to a high number of (short) network chains. The proportionality factor $\langle r^2\rangle_u/\langle r^2\rangle_0$ in eq. (5.66) describes the ratio of the mean-square end-to-end distance of the *network chains* in the undeformed state to the mean-square end-to-end distance of an analog *uncrosslinked ideal chain.* This factor depends on the state of the network chains during the network formation. Often, the network is made in a state where the chains that are crosslinked (or formed in-situ during crosslinking) are in the same state (e.g., an ideal or θ-state) as they are during the network deformation experiment; in this case, the factor $\langle r^2\rangle_u/\langle r^2\rangle_0$ in eq. (5.66) is one. It may deviate from one when the states of network preparation and network probing are different, for example, when the network was formed in a swollen state but is measured in a deswollen state, or vice versa. For the rest of our argumentation, we assume the usual case of $\langle r^2\rangle_u = \langle r^2\rangle_0$.

Now that we have derived an expression for the change of entropy, ΔS, we can insert it into the fundamental equation for the free energy, eq. (5.56), and obtain

$$\Delta F = \Delta U - T\Delta S = +\frac{nk_BT}{2}\left(\lambda_x{}^2 + \lambda_y{}^2 + \lambda_z{}^2 - 3\right) \quad (5.67)$$

The internal energy, U, is independent of the network-chain end-to-end distance, as we presume these chains to be ideal, such that $\Delta U = 0$. Any change of the network chains' free energy, ΔF, therefore stems from entropy alone.

Let us consider a typical experimental situation: isochoric uniaxial strain. For that situation, the latter expression can be simplified further. Isochoric means that the total volume change upon deformation is zero, $\Delta V = 0$, which means that the product of the strain ratios has to be one, $\lambda_x\lambda_y\lambda_z = 1$. Uniaxial means that the strain is only applied along one spatial dimension, x, leading to one defined strain ratio $\lambda_x = \lambda$, whereas the two other strain ratios must follow according to the isochoric condition: $\lambda_y = \lambda_z = 1/\sqrt{\lambda}$.

Both considerations lead to a simplified expression:

$$\Delta F = \frac{nk_BT}{2}\left(\lambda^2 + \frac{2}{\lambda} - 3\right) \quad (5.68)$$

The deformative force f_x can again be calculated by the length derivate along the axis of deformation:

$$f_x = \frac{\partial \Delta F}{\partial L_x} = \frac{\partial \Delta F}{\partial(\lambda L_{x,0})} = \frac{nk_BT}{L_{x,0}}\left(\lambda - \frac{1}{\lambda^2}\right) \quad (5.69)$$

By normalizing it to the plane of deformation, we can formulate an expression for the stress σ_{xx}

$$\sigma_{xx} = \frac{f_x}{L_yL_z} = \frac{nk_BT}{L_{x,0}L_yL_z}\left(\lambda - \frac{1}{\lambda^2}\right) = \frac{nk_BT}{L_{x,0}\lambda_yL_{y,0}\lambda_zL_{z,0}}\left(\lambda - \frac{1}{\lambda^2}\right) \quad (5.70a)$$

Due to the isochoric condition, $\lambda_y = \lambda_z = 1/\sqrt{\lambda}$, we can eliminate the direction-specific strain ratios and simplify the expression to

$$\sigma_{xx} = \frac{nk_BT}{L_{x,0}L_{y,0}L_{z,0}}\lambda\left(\lambda - \frac{1}{\lambda^2}\right) = \frac{nk_BT}{V}\left(\lambda^2 - \frac{1}{\lambda}\right) \quad (5.70b)$$

In the very beginning of our discussion on rheology, in eq. (5.1), we have defined the extensional strain as $\varepsilon = \Delta L/L_0$, that is, the change of the material specimen length, ΔL, relative to its length before deformation, L_0; for uniaxial strain, this is equal to $\varepsilon = \lambda - 1$, so $\lambda^2 - \frac{1}{\lambda} \approx \varepsilon^2 + 2\varepsilon + 1 - (1 - \varepsilon) = \varepsilon^2 + 3\varepsilon$,[61] which for small ε is nearly 3ε. With that, we get

$$\sigma_{xx} = \frac{nk_BT}{V} 3\varepsilon \tag{5.70c}$$

This equation is strikingly similar to Hooke's law, eq. (5.1), that connects stress to strain by a proportionality factor. In eq. (5.70c), this factor is $3\,\frac{nk_BT}{V}$; it therefore corresponds to the elastic modulus:

$$E = 3\frac{nk_BT}{V} \tag{5.71a}$$

From that, along with eq. (5.3), we also get the shear modulus:

$$G = \frac{E}{3} = \frac{nk_BT}{V} \tag{5.71b}$$

The latter formula is quite an extraordinary expression, as it connects the microscopic structural information of the network chain concentration, n/V, to the macroscopic property of the shear modulus, G, thereby creating a quantitative structure–property relation. We now realize that upon deformation, **each network chain stores an energy increment of k_BT.** The modulus increases with temperature due to the entropy elastic nature of rubber elasticity. We also realize that a higher degree of crosslinking causes a higher modulus, as the network chains are then more numerous (and shorter), leading to a higher n in eq. (5.71a).

We can formulate the latter identity on a molar scale as well:

$$G = \frac{\rho RT}{M_x} \tag{5.71c}$$

This equation relates the network density or network concentration, ρ, in the unit $g \cdot L^{-1}$, to the molar mass of the network chains, M_x, in the unit $g \cdot mol^{-1}$. From this, we again realize that a higher degree of crosslinking causes a higher modulus, as the network chains are then shorter (and more numerous), leading to a lower M_x in the denominator of eq. (5.71b).

61 The calculation for this is as follows: $\lambda - \frac{1}{\lambda^2} = (\varepsilon + 1)^2 - \frac{1}{\varepsilon+1} = \frac{(\varepsilon+1)^3}{\varepsilon+1} - \frac{1}{\varepsilon+1} = \frac{(\varepsilon+1)^3 - 1}{\varepsilon+1} = \frac{\varepsilon^3 + 3\varepsilon^2 + 3\varepsilon}{\varepsilon+1}$. Polynomial long division turns that into $\varepsilon^2 + 2\varepsilon + 1 - \frac{1}{\varepsilon+1}$. Taylor series approximation of $\frac{1}{\varepsilon+1} \approx 1 - \varepsilon$ (as first described by Isaac Newton) then gives $\varepsilon^2 + 2\varepsilon + 1 - (1 - \varepsilon)$.

Equation (5.71b) is strikingly similar to the ideal gas law in elementary physical chemistry. This is because rubber elasticity at all is very similar to the concept of pressure for an ideal gas. Both are based on the reduction of freedom of arrangement for the systems building blocks (gas particles in space, or monomer units in a polymer chain) upon application of deformation such as compression of the ideal gas, which limits the freedom of spatial arrangement of the gas particles, or stretching of the polymer chain, which limits the freedom of arrangement of the monomers along the chain from a mix of *cis* and *trans* local conformation to more *trans*. We can actually use the same statistical thermodynamic approach that we have used above to appraise the pressure of an ideal gas: we appraise the probability, Ω, for a gas molecule to sit in a subvolume V of a container with total volume V_0 to be $\Omega = V/V_0$. From this follows the fact that the probability that n gas molecules *all* sit in that subvolume at a time is $\Omega_n = (V/V_0)^n$. We estimate the system's entropy, S, by inserting this into the Boltzmann formula: $S = k_B \ln\Omega_n = k_B n \ln(V/V_0)$. From physical chemistry lectures, we know that the pressure, p, in thermodynamics is calculated as $p = -(\partial G/\partial V)_T = -(\partial H/\partial V)_T + T(\partial S/\partial V)_T$. In our case, we are looking at an ideal gas, whose particles do not interact. This means that there is no energetic change upon change of volume, that is, $(\partial H/\partial V) = 0$. This simplifies the latter expression to $p = T(\partial S/\partial V)_T = (k_B n \cdot T)/V$. This is exactly equal to eq. (5.71a).[62]

An alternative calculation can be made for an isotropic shear experiment. In this case, there is no deformation in the z-direction; hence $\lambda_z = 1$. The two other strain ratios are then defined to be $\lambda_x = \lambda$ and $\lambda_y = 1/\lambda$. For the free energy, this yields

$$\Delta F = \frac{nk_BT}{2}\left(\lambda^2 + \lambda^{-2} - 2\right) = \frac{nk_BT}{2}\left(\lambda - \lambda^{-1}\right)^2 = \frac{nk_BT}{2}\gamma^2 \tag{5.72}$$

Subsequent calculation of the stress gives

$$\sigma = \frac{nk_BT}{V}\gamma \tag{5.73}$$

Again, we end up with an expression that has the form of Hooke's law. In this case, we identify the proportionality factor nk_BT/V to be the shear modulus G:

$$G = \frac{nk_BT}{V} \tag{5.74a}$$

Once again, we can formulate the latter identity on a molar scale as well:

$$G = \frac{\rho RT}{M_x} \tag{5.74b}$$

62 Note from this identity that pressure and the elastic modulus also have the same unit: Pascal.

So far, we have discussed rubber elasticity on the basis of permanent crosslinks. Another form of hindering polymer chains from relaxing, at least temporarily, is mutual **entanglement**, as shown in Figure 91(A). Hence, on intermediate timescales, such entanglements act like permanent crosslinks and create a rubbery plateau in the time- or frequency-dependent elastic modulus, as shown in Figure 91(B).

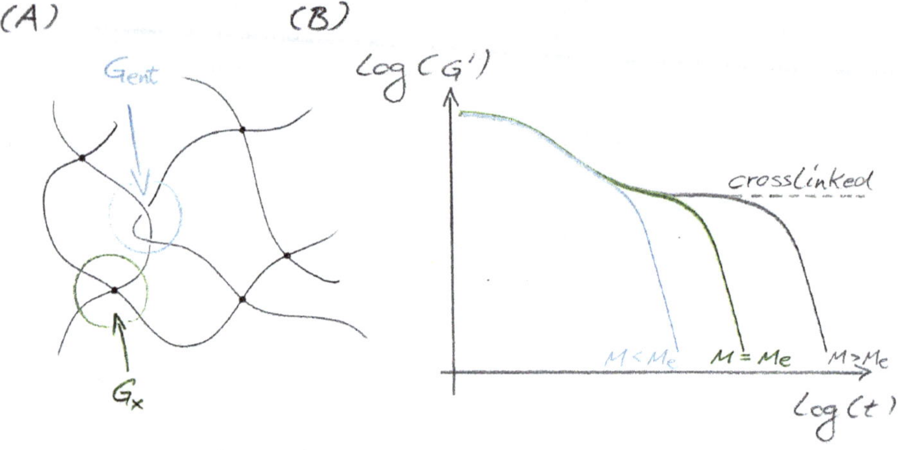

Figure 91: **(A)** Schematic representation of a polymer network composed of permanent crosslinks and chain entanglements. Both hinder the chains from relaxing and therefore contribute to the ability of elastic energy storage, thereby both contributing to the plateau modulus in the form of $G_p \cong G_x + G_{ent}$. For permanently crosslinked networks, the plateau in the mechanical spectra, shown in **(B)**, is infinite and reaches out to $t \to \infty$, whereas if no permanent crosslinks are present, the polymer chains can untie their mutual entanglement at long times. The extent of the rubbery plateau is defined by the polymer molar mass, which must exceed a specific minimum molar mass, M_e, to show mutual entanglement at all.

A crosslinked network in fact usually has both modes of polymer connection: crosslinking and (trapped) entanglements, as shown in Figure 91(A). We can therefore split the modulus G into two contributions, one from actual permanent crosslinks, G_x, and one from (trapped) entanglements, G_{ent}:

$$G \cong G_x + G_{ent} = \rho RT\left(\frac{1}{M_x} + \frac{1}{M_{ent}}\right) \tag{5.75}$$

This means that even uncrosslinked polymer systems, in which $G_x = 0$, are elastic on short timescales solely due their mutual entanglements, whereby G_{ent} typically is in the range of a megapascal:

$$G_{ent} = \frac{\rho RT}{M_{ent}} \approx 10^6 \text{ Pa} \tag{5.76}$$

There is just one prerequisite for this: to entangle with one another, chains must have a certain minimum length, which translates to a certain minimum molar mass, named the **entanglement molar mass**, M_e. This quantity is related to the one found in the denominator of the last equations, named M_{ent} there. Typically, $M_e \approx 2\ M_{ent}$, because M_e denotes the minimum molar mass (and therewith also the minimum degree of polymerization N_e, and yet in turn, the minimum contour length $N_e l$) that chains must have to form entanglements, whereas M_{ent} denotes the length of a strand segment between two entanglement points in a well-entanglement network. A simple geometrical thought leads us to the notion that M_e must be twice as large as M_{ent}, because to form entanglements at all, the chain must be minimally twice as long as one segment between two entanglement nodes in an entangled network.

The reason for the existence of a minimum molar mass for entanglement is because the chains have a certain stiffness or persistence, which must be overcome by sufficient chain length to form entanglements. There is an analogy to everyday life. In large canteens, for example, in the Mensa, when spaghetti are cooked and served, they are short. This is because the kitchen personnel wants the noodles to be shorter than their entanglement length, as entangled noodles are much harder to process than unentangled ones. The stiffer the main chain is (which, on the nanoscale, translates to the ease or non-ease of monomer-bond rotatability), the longer must it be to be able to form entanglements. For polystyrene, an entanglement molar mass of $M_e \approx 20\ \text{kg mol}^{-1}$ is practically relevant.

As a closing note, it shall be mentioned that all the above appraisal is based on the entropy of an ideal chain, which we have obtained in Section 2.6.1 by applying Boltzmann's formula to the distribution of end-to-end distances that we have derived by random-walk statistics before. Hence, all the above applies to chains only with these statistics. This is valid only in the limit of low deformation, whereas at large deformation, we actually disturb the random-walk-type distribution of end-to-end distances considerably, such that it is no longer valid. In this limit, other statistics have to be used, for example, the Langevin model.[63] In addition, in polymers with high main-chain regularity and tacticity, crystallization is possible. Stretching of the network chains may then bring them in so close distance and good order that crystallization may occur, leading to crystalline nodes that act as further crosslinks. This phenomenon is named strain-induced hardening.[64]

---

63 Again, there is an analogy to ideal gases here: at large pressure, they deviate from ideality and must be treated by a different model: the van der Waals equation.

64 And once more, there is an analogy to gases, which may liquefy at high pressure.

5.9.3 Swelling of rubber networks

In addition to stretching or shearing, there is another way to deform a polymer network: **swelling**. In the process of swelling, a fluid medium enters the network and expands it from the interior. The basis of this process is that upon contact of solvent molecules with a dry network, its chains want to dissolve. Their mutual connectivity, however, hinders them to do so freely. Hence, rather than separating the chains from one another completely, as it would be the case in an uncrosslinked sample, solvent can only penetrate into the network and stretch the polymer network chains in it. The degree of stretching, however, is limited by the entropy-elastic backdriving force that comes along with that, as we have just quantified above. At equilibrium, the polymer–solvent mixing energy is at balance with the entropy-elastic energy. The point where this equilibrium lies, and with that, the **swelling capacity** of the network, depends on the ratio of the network chain length (or in other words, the density of crosslinking points in the network) and the thermodynamic quality of the solvent (athermal, good, or Θ).

To quantify the swelling equilibrium, we introduce the **degree of swelling** q as

$$q = \frac{V}{V_0} = 1 + \frac{V_{\text{uptaken solvent}}}{V_{\text{unswollen network}}} \tag{5.77}$$

To estimate this value, we need to know both contributors: the free energy of mixing and the entropy-elastic backdriving energy. This is best gauged via the *chemical potentials* related to these energies.

The contribution of the free energy of mixing is given by the Flory–Huggins formula:

$$\Delta\mu_{\text{mix}} = RT\left(\ln\left(1 - q^{-1}\right) + q^{-1} + \chi q^{-2}\right) \tag{5.78}$$

The potential of the entropy-elastic backdriving force is calculated by

$$\Delta\mu_{\text{elast}} = \left(\frac{\partial \Delta F_{\text{elast}}}{\partial n}\right) = \left(\frac{\partial \Delta F_{\text{elast}}}{\partial q}\right) \cdot \left(\frac{\partial q}{\partial n}\right) \tag{5.79}$$

The elastic energy is given by the rubber elasticity formula:

$$\Delta F_{\text{elast}} = \frac{3}{2} N k_B T \left(q^{2/3} - 1\right) \tag{5.80}$$

Here, N denotes the *total number* of network chains.

Assuming that both the volumes of the dry network and the solvent are additive, we can formulate the degree of swelling as follows:

$$V = n_{\text{solvent}}\,\bar{V}_{\text{solvent}} + V_0 \Rightarrow q = \frac{V}{V_0} = n_{\text{solvent}}\frac{\bar{V}_{\text{solvent}}}{V_0} + 1 \Rightarrow \left(\frac{\partial q}{\partial n}\right) = \frac{\bar{V}_{\text{solvent}}}{V_0}$$

$$\Rightarrow \Delta\mu_{\text{elast}} = \left(\frac{\partial \Delta F_{\text{elast}}}{\partial q}\right)\cdot\left(\frac{\partial q}{\partial n}\right) = \frac{3}{2}Nk_{\text{B}}T\cdot\frac{2}{3}q^{-1/3}\cdot\frac{\bar{V}_{\text{solvent}}}{V_0} = vRTq^{-1/3}\cdot\bar{V}_{\text{solvent}} \quad (5.81)$$

Here, v denotes the *concentration of network chains*, which is inversely related to the network chain length. This is because in a network with a given amount of polymer material in it, if the network chains are long, then there is just few of them and vice versa. Hence, v can be considered to be both the molar number of network chains n per volume unit V [mol·L^{-1}], or the network density ρ [g·L^{-1}] divided by the network chain molar mass M_x [g·mol^{-1}]:

$$v = \frac{n}{V} = \frac{\rho}{M_x} \quad (5.82)$$

At the swelling equilibrium, both the free energy of polymer–solvent mixing as well as the entropy-elastic backdriving energy are identical:

$$\Delta\mu_{\text{mix}} = \Delta\mu_{\text{elast}} \quad (5.83)$$

This allows us to calculate the molar concentration of network chains as

$$v = \frac{\rho}{M_x} = -\frac{\ln(1-q^{-1}) + q^{-1} + \chi q^{-2}}{\bar{V}_{\text{solvent}}\,q^{-1/3}} \quad (5.84)$$

We can now understand the value of doing a swelling experiment. It allows us to determine the degree of crosslinking at a known Flory–Huggins parameter χ, or, vice versa, the determination of the Flory–Huggins parameter at a known degree of crosslinking.

The swelling ability and often-soft appearance of polymer gels give rise to various areas of application. Particularly relevant in this context are hydrogels, which are polymer networks swollen with up to 99% (w/w) of water. This is, in principle, the same ratio of water and solid material as in biological matter, which is why hydrogels are often used as matrixes in the field of life science for procedures such as electrophoresis or chromatography. Dried hydrogels can also be found in hygiene products as super absorbers due to their superior ability to absorb moisture.

In more sophisticated applications, the polymer network can be designed to react to external stimuli that trigger either swelling or shrinking. Such a stimulus-sensitive reversible swelling is sketched in Figure 92. Stimuli-responsive gels have potential uses as artificial muscles, chemo-mechanical actuators, or as "intelligent" capsules for drug release. As an example for the latter, consider the following: cancer tissue has a marginally lower pH than the surrounding healthy tissue. In a specifically tailored polymer network, the pH change would trigger a deswelling or a

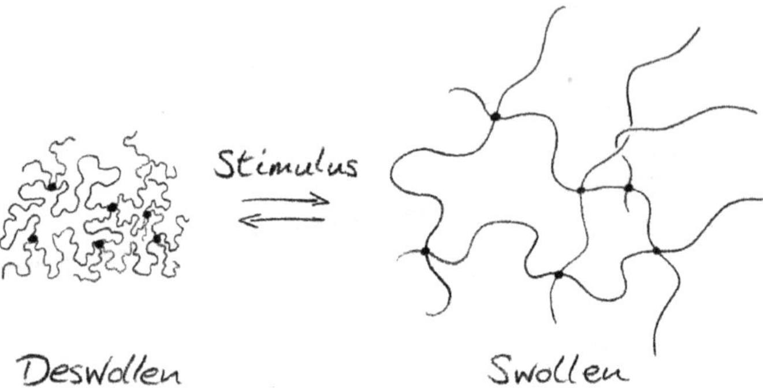

Figure 92: Schematical representation of a stimulus-sensitive polymer network that can undergo swelling or deswelling upon external triggers. This stimuli-responsiveness renders such materials useful for potential applications such as chemo-mechanical switches, artificial muscles, or "smart" drug-release carriers.

swelling process, thus initiating controlled release of an cytostatic drug (either by squeezing it out of the gel in the case of deswelling or by letting it diffuse out of the gel upon opening diffusive paths through the widened network meshes in the case of swelling). The same principle might be used for the controlled release of anti-inflammatory agents. Inflamed tissue has a higher local temperature than its environment; hence, in that case, the temperature difference could provide a stimulus for the hydrogel to shrink or swell.

5.10 Terminal flow and reptation

LESSON 16: REPTATION

When a noncrosslinked polymer sample is given enough time, it will display viscous flow, even if the chains are long and entangled. On a microscopic scale, such flow requires the chains to change their positions against each other. Describing this microscopic motion is an essence to describe macroscopic material properties such as the polymer fluid viscosity. In this lesson, you will get to know a simple but powerful approach to do so: the reptation model. This model simplifies the direct environment of a chain in an entangled surrounding to be a rigid tube, out of which the chain can creep only by curvilinear motion.

So far, our knowledge on the viscoelastic spectra of polymers has two sides: a phenomenological macroscopic one and a conceptual microscopic one. On the one side, phenomenologically, we can describe the macroscopic viscoelastic solid-like mechanics of the short-time or high-frequency regime with the Kelvin–Voigt model, which gives us equations to approximate the leather-like creep properties of a sample in that domain. As a complement to that, we can use the Maxwell model to describe the viscoelastic liquid-like stress relaxation of the sample in the long-time or low-frequency domain. On the other side, conceptually and microscopically, we can understand how the viscoelastic solid-like appearance of the sample in the first domain originates from viscoelastic relaxation modes, that is, motion of only parts of the chains but not yet the chains as a whole, which we can quantify based on the Rouse or the Zimm model. We also understand how chain entanglement may introduce a further domain into the mechanical spectra, namely, that of a rubbery plateau. What we are lacking, though, is a conceptual microscopic understanding of the relaxation and flow in the terminal regime. Delivering that is the intent of the following section. We have already seen that the transition between the rubbery plateau and the terminal flow regime depends on something called the *entanglement molar mass*, M_e (see Figure 91(B)), without further discussing this quantity; we will do so in this subchapter.

Describing the long-term relaxation of an entangled multichain system is a challenging proposition. Many chains move simultaneously through the sample volume, part in concert with each other, part independently. They may also interact with each other or, if present, with a solvent. This is a very complex multibody phenomenon that can hardly be described analytically. The solution to this challenge is as easy as it is clever: we focus only on one **test chain**, and then simplify its environment drastically.

5.10.1 The tube concept

Consider a test chain in an ensemble of overlapping and entangled chains that we have somehow modified so we can identify it from the background and be able to follow its path of motion, as sketched in Figure 93(A). To describe its surroundings, we neglect every other chain that does not have contact with our test chain. Even the ones that have contact only do so with small portions of their own length, such that we must only regard these sections of them, as illustrated in Figure 93(B). These contacts form an array of obstacles in the direct environment of our test chain. Now, a smart approach by Samuel Frederick Edwards comes up. Edwards modeled the direct constraining environment of the test chain by a solid tube, as shown in Figure 93(C). The test chain can move only along this tube in a curvilinear manner, whereas perpendicular movement to the tube is prohibited. With this kind of motion, the chain can creep out of its confining tube, as illustrated in Figure 93(D).

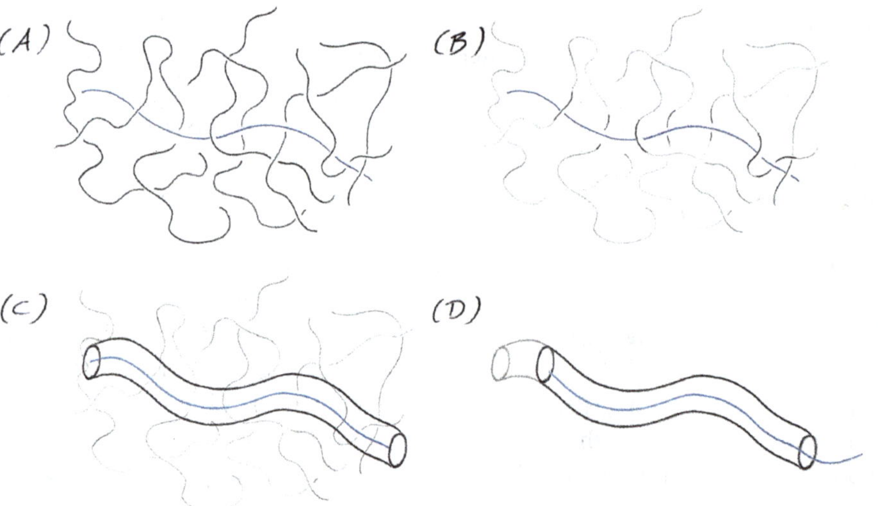

Figure 93: Tube concept to simplify the complex multibody surrounding of a polymer chain in a system of many other mutually entangled chains. **(A)** We consider a test chain, here labeled in blue color, which is somehow distinguishable from its background into which it is embedded and with which it is entangled. **(B)** To simplify our view, we only consider the direct environment of the test chain. **(C)** The directly surrounding matrix chain segments around the test chain can be modeled as a confining solid tube that constrains the motion of the test chain to be possible only along the tube contour. **(D)** Over time, the test chain can creep out of this confining tube by curvilinear motion, which is named *reptation*. Note: as the surrounding matrix is still there, the test chain then finds itself entrapped in a new tube (not shown here). The overall diffusion of the test chain through the surrounding matrix can be viewed as a sequence of steps from one tube to another, each of which occurring by curvilinear creep along the tube contour. Pictures inspired by W. W. Graessley: Entangled linear, branched and network polymer systems – Molecular theories, *Adv. Polym. Sci.* **1982**, *47* (Synthesis and Degradation Rheology and Extrusion), 67–117.

This approach is called the **tube concept**. Mathematically, Edwards assumed that each neighboring segment imparts a parabolic confining potential on the test chain. The energetically most favorable way for the test chain to arrange itself in the tube and also to move along it is through the path of the potential minima through space, which is called the **primitive path**. The test chain fluctuates around this primitive path, but only up to an extent that the ever-present thermal energy $k_B T$ can activate, because anything further would require additional energy input. The **tube diameter**, a, is therefore given by the transverse distance at which the confining potential is exactly $k_B T$. In other words, segments of the test chain fluctuate with an average displacement of a.

5.10.2 Rouse relaxation and reptation

In the tube concept, the motion of a chain is largely restricted only to curvilinear creep along the tube contour, whereas migration in directions perpendicular to that is impossible. Pierre-Gilles de Gennes used this premise to mathematically describe the test chain motion along the tube, which, as it resembles the creep of a reptile through the woods, is referred to as **reptation**. De Gennes based his approach upon the Rouse model for polymer dynamics in the melt, in which the diffusion along the tube would be described by a one-dimensional Rouse-type diffusion coefficient:

$$D_{\text{Tube}} = \frac{k_B T}{N f_{\text{seg}}} \tag{5.85}$$

The diffusion of a polymer chain is related to its mean-square displacement by the Einstein–Smoluchowski equation that we have encountered numerous times in Section 3.6: $\langle x^2 \rangle = 2dDt$, with d the geometrical dimension of the motion. We can now adopt this for the reptating polymer: the tube has a contour length equal to the total length of the polymer test chain, $L = Nl$, which will leave the tube after a specific time τ_{rep}, the **reptation time**:

$$\tau_{\text{rep}} = \frac{\langle L_{\text{Tube}}{}^2 \rangle}{2 \cdot D_{\text{Tube}}} = \frac{N^2 l^2}{2 k_B T / N f_{\text{seg}}} \sim N^3 \tag{5.86}$$

We see that the degree of polymerization, and with that, the molar mass of the test chain, has a paramount influence on the reptation time, as it scales with a power-law exponent of 3(!).

Furthermore, we have learned about the fundamental connection between the shear modulus, G, and the viscosity, η, by the relaxation time, τ: $\eta = G_0 \tau$ (eq. 5.46). To estimate the viscosity, we only need to plug in the relevant values:

In *nonentangled melts*, the relaxation is delimited by the Rouse time, τ_{Rouse}. Under melt conditions, the chain conformation is ideal, and in that case, the Flory exponent is $\nu = ½$, such that we get from the Rouse model (cf. Section 3.6.2):

$$\tau_{Rouse} = \tau_0 N^{1+2\nu} = \tau_0 N^2 \tag{5.87}$$

The latter equation features two characteristic timescales: τ_0, which denotes the lower limit for any chain motion at all (below that timescale, not even single monomer segments can get displaced by a distance at least equal to their own size), and τ_{Rouse}, which denotes the upper time limit for complete chain motion (above that timescale, the whole coil can get displaced by distances equal to or even greater than its own size). In the time domain between these two characteristic limits, the modulus G relaxes from an energy-storage capacity of $k_B T$ per monomer segment at τ_0 to an energy-storage capacity of $k_B T$ per chain at τ_{Rouse} (see Section 5.8.2). The shear modulus at the Rouse time is therefore given by

$$G(\tau_{Rouse}) = k_B T \frac{\phi}{Nl^3} \tag{5.88}$$

Here, ϕ is the polymer volume fraction, and Nl^3 is the volume per chain in a fully collapsed state, such that ϕ/Nl^3 is an expression for the number of chains per volume in the sample, denoted ν in Section 5.8.2. With that, the modulus at τ_{Rouse} is $k_B T$ per chain in the sample volume.

According to eq. (5.46), the viscosity at the Rouse time can then be calculated as

$$\eta = G(\tau_{Rouse}) \cdot \tau_{Rouse} \sim N^{-1} N^2 \sim N^1 \tag{5.89}$$

We realize that it scales linearly (power-law scaling exponent of 1) with the degree of polymerization, and with that, with the molar mass.

The picture is very different for *entangled melts*. Here, the relaxation is delimited by the reptation time, τ_{rep}

$$\tau_{rep} \sim N^3 \tag{5.90}$$

At τ_{rep}, we find ourselves just at the end of the rubbery plateau of the shear modulus G. Here, the modulus just depends on the entanglement molar mass, M_e, (see eq. (5.76)), but *not* on he overall molar mass, M_{total}. This leads to a different dependence of the viscosity on the molar mass, and with that, on the degree of polymerization N, than that of eq. (5.89):

$$\eta = G(\tau_{rep}) \cdot \tau_{rep} \sim \text{const} \cdot N^3 \tag{5.91}$$

In eqs. (5.89) and (5.91), we have again connected the macroscopic properties viscosity, η, and shear modulus, G, to the microscopic property relaxation time, τ, that contains information about the molecular-scale characteristics of the system's building blocks. In particular, we were able to delimit two very different viscosity–molar-mass scaling laws, depending on whether a specific molar mass, the entanglement molar mass, is surpassed. These dependencies are visualized in Figure 94.

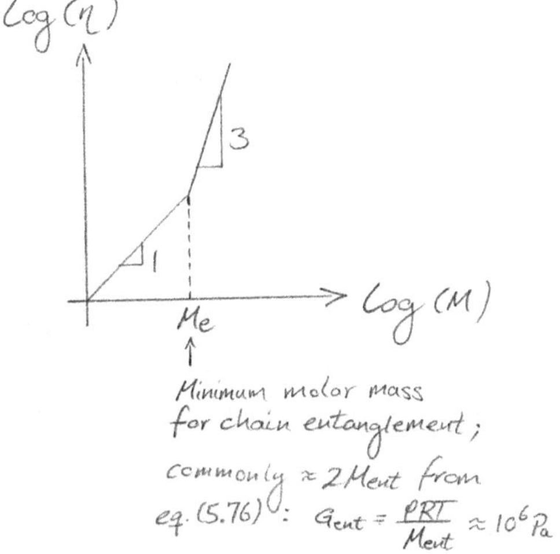

Figure 94: Plot of a polymer melt sample's viscosity as a function of the chains' molar mass. Up to certain molar mass, M_e, the dependence is linear. Once this molar mass is exceeded, however, the chains entangle with each other. This imparts severe topological constraint on the polymer chains' motion, which is now to be described by P.G. de Gennes' reptation model. Here, the viscosity has a severe molar-mass dependence with a power-law exponent of 3.

But how come that we need a minimal M_e for chain entanglement? The reason is the polymer chain stiffness, as quantified by parameters such as the characteristic ratio, the Kuhn length, or the persistence length, as all introduced in Chapter 2. The stiffness of the chains requires them to have at least a certain length to bend enough to form entanglements, which is longer if the chains are stiffer. An every-day life analogy for the phenomenon of a minimum length for entanglement is spaghetti: when being served in a proper Italian restaurant, they are long and entangle with one another quite markedly. By contrast, when being served in a large canteen, for example, a university Mensa, they are short such that they can *not* entangle. The kitchen personnel does that to avoid processing problems of entangled noodles. So, even spaghetti have a minimum length (or "molar mass") for entanglement.

Getting back to Figure 94: note that experimental data actually suggest an $\eta(M)$ scaling-law exponent of 3.4 rather than 3 for the initial reptation regime. This is due to fluctuations of the length of the confining tube. The chain ends fluctuate on timescales $t < \tau_{rep}$. Because of this fluctuation, the effective tube length is actually shorter, which decreases τ_{rep}. This effect, however, gets less significant when the molar mass M increases. Hence, in addition to the primary effect of an M-increase on η, as captured by eq. (5.90), comes a secondary one by the no-longer-effectiveness of tube-length fluctuation. Together, this leads to the observed stronger molar-mass-dependence of a power of 3.4. In the very high molar-mass regime, the effect of such tube length fluctuation looses significance, such that the original reptation power-law exponent of 3 is recovered there.

5.10.3 Reptation and diffusion

In addition to discussing the macroscopic viscosity of our sample, we may also quantify a microscopic characteristic parameter of it: the translational diffusion coefficient of the chains. On timescales shorter than τ_{rep}, the chains exhibit a **one-dimensional tube diffusion coefficient, D_{Tube},** as captured by eq. (5.85). On timescales longer than τ_{rep}, by contrast, each chain diffuses three-dimensionally through space with a **Fickian diffusion coefficient D_{rep}.** We can imagine this diffusion to be a hopping process from tube to tube, each of these hops taking a time increment of τ_{rep}. Based on that picture, the macroscopic chain displacement in a three-dimensional space can again be quantified by an Einstein–Smoluchowski equation:

$$D_{rep} \approx \frac{R^2}{\tau_{rep}} \sim \frac{Nl^2}{N^3} \sim N^{-2} \tag{5.92}$$

This power-law scaling has indeed been well confirmed by multiple sorts of experiments.

When we focus on the microscopic diffusion of polymer chains in a reptation scenario, as we have just started to do above by quantifying the two relevant chain diffusion coefficients, D_{tube} and D_{rep}, we can also take an even more detailed view. In that detailed view, we consider one monomer segment on a reptating chain to be somehow labeled, for example, by radioactive marking or fluorescent tagging. When we follow the mean-square displacement, $\langle \Delta r^2 \rangle$, over time, t, of that labeled segment, we can distinguish four different regimes, as shown in Figure 95(B).

1) On timescales between the segmental relaxation time, τ_0, below which no segmental motion at all is possible, up to the entanglement time, τ_e, from which on the segments first realize that they are trapped in a tube environment, the segmental displacement is impaired by the connectivity to neighboring segments. This constraint gives us a time dependence of the mean-square displacement according to the Rouse model:

$$\langle \Delta r^2 \rangle \sim t^{1/2} \tag{5.93}$$

2) At the entanglement time, τ_e, the chain segment also takes notice of the additional confining tube. Up to the Rouse time, τ_{Rouse}, the segmental displacement along the tube is still Rouse type, but in three dimensions it is

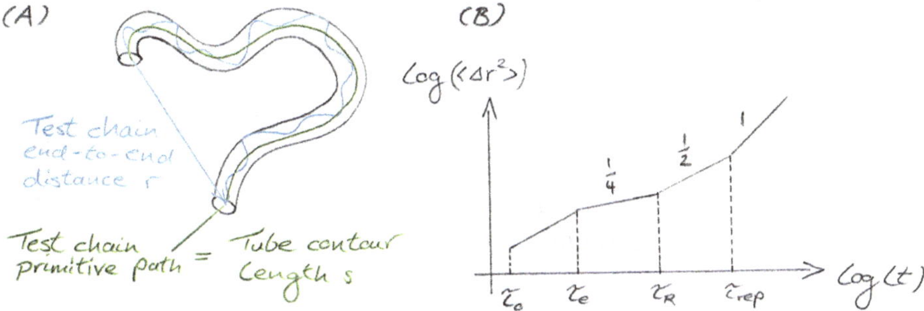

Figure 95: **(A)** Visualization of a chain entrapped in a confining tube with end-to-end distance *r* and contour length *s*. **(B)** Subdiffusive displacement of a labeled chain segment as a function of time on different timescales during a reptation process. The mean-square displacement, $\langle\Delta r^2\rangle$, of the labeled segment is proportional to time, *t*, only on scales longer than τ_{rep}, because only on these long timescales, the labeled segment follows the three-dimensional Fickian diffusion of the chain through space (which we may imagine as a hopping process from one tube to another with an overall three-dimensional diffusion coefficient D_{Rep}). On shorter scales, by contrast, the motion of the chain segment under consideration is sub-diffusive. First, on very short timescales, we have subdiffusive spreading of the labeled segment due to the constraint imparted by its connectivity to neighbor segments, which gives a time dependence of $\langle\Delta r^2\rangle\sim t^{1/2}$ according to the Rouse model (which applies in polymer melts). Second, on longer timescales, from the entanglement time τ_e on, additional constraint comes into play, as our labeled segment now realizes that it is also entrapped in a tube and can only move along the tube contour, which in 3D space is a random walk with scaling of $r\sim N^{1/2}$ such that the overall scaling exponent is ½ · ½ = ¼. Third, on even longer timescales, from the Rouse time τ_R on, the first constraint is lost, but the second constraint is still active, such that we still have subdiffusive scaling, now again according to $\langle\Delta r^2\rangle\sim t^{1/2}$. Fourth, only on timescales longer than the reptation time, τ_{Rep}, also the latter constraint is lost and the chain moves freely in space, and so does the labeled segment on it.

$$\langle\Delta r^2\rangle \sim \sqrt{\langle s^2\rangle} \sim \sqrt{t^{1/2}} \sim t^{1/4} \tag{5.94}$$

This is because the tube itself has a random walk shape with $r\sim s^{1/2}$. In other words, in the time window τ_e–τ_{Rouse}, our labeled segment is constrained by both its connectivity to other segments, giving us one scaling-exponent contribution of ½ according to eq. (5.29), and on top of that, it now also realizes that it is entrapped in a tube and can only move along the tube contour, which in three-dimensional space is a random walk with scaling of $\langle r^2\rangle\sim\sqrt{\langle s^2\rangle}$, giving us another scaling-exponent contribution of ½, such that the overall scaling exponent is ½ · ½ = ¼.

3) At times longer than the Rouse time, τ_{Rouse}, we now discuss the motion of the *entire* coil along the tube. Its mechanism of motion is still a one-dimensional Rouse-type diffusion, but in three dimensions it is

$$\langle\Delta r^2\rangle \sim \sqrt{\langle s^2\rangle} \sim \sqrt{D_{Tube}t} \sim t^{1/2} \tag{5.95}$$

4) On timescales longer than the reptation time, $t > \tau_{rep}$, we have macroscopic three-dimensional diffusion of the entire coil according to the Einstein–Smoluchowski equation:

$$\langle \Delta r^2 \rangle \sim 6 D_{rep} t \tag{5.96}$$

The polymer chain now has enough time to "hop" from one confining tube to another and thereby move freely through space.

5.10.4 Constraint release

In the years after development of the reptation model, several refinements have been added to it to better match to experimental data. An impactful refinement was the discovery of the so-called **constraint release mechanism.** It takes into account that, in addition to the test chain, all other chains of the system are in motion as well. As a consequence, a topological constraint resulting from an entanglement point might resolve itself by diffusion of the constraining chain, as depicted in Figure 96. The effective diffusion is, thus, determined by reptation and the finite lifetime of the momentary tube:

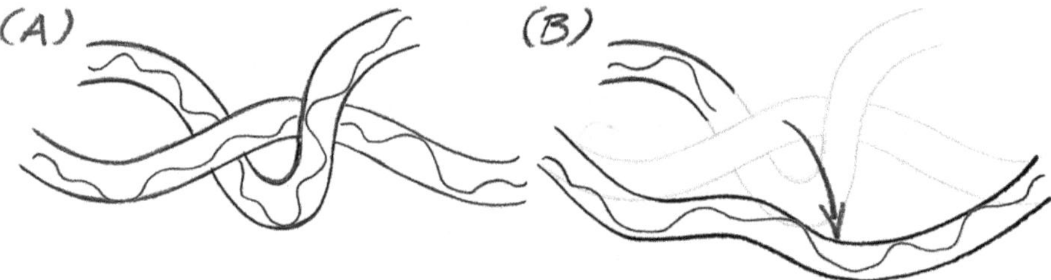

Figure 96: Constraint release as an additional mechanism for chain relaxation in entangled systems. (**A**) Schematic of two entangled chains entrapped in their respective tubes. (**B**) Reptation of one chain opens a degree of freedom for the other chain to rearrange itself not only by curvilinear motion along its tube contour, but also lateral to it. Picture redrawn from M. Rubinstein, R. H. Colby: *Polymer Physics*, Oxford University Press, **2003**.

$$D_{eff} \approx \frac{R^2}{\tau_{rep}} + \frac{R^2}{\tau_{Tube}} \tag{5.97}$$

Depending on which process is faster, the overall diffusion is dominated by reptational motion or the constraint release mechanism. As a consequence, both the tracer and matrix molar mass are important.

6 Scattering analysis of polymer systems

Lesson 17: Scattering Methods in Polymer Science

The overall goal of polymer physical chemistry is to bridge between polymer structures and properties. This requires both to be assessed. In this lesson, you will get to know scattering methods as a prime experimental approach to assess polymer structures on different scales. You will learn about the scattering vector as the "magnifying glass" in scattering, and you will see how the scattering intensity as a function of this vector is connected to microstructural aspects.

6.1 Basics of scattering

The basic target of this book – and of the field of physical chemistry of polymers in general – is to unravel relations between structure and properties of polymer systems. In Chapter 5, we have learned that the most relevant class of properties of polymers are their mechanical characteristics, and we have learned about a class of methods to quantify these: rheology. What we are lacking still is an equivalent class of methods to characterize polymer *structures*. We know that the size of polymers and their superstructures is in the colloidal domain of 10...1000 nm, so the methodology for structural characterization that we seek for must be able to resolve that. Most polymer systems are amorphous, while only a few can show regular crystalline structures. Hence, we also need a methodology that is generally able to assess both in the colloidal size range. A class of methods that serves all that is based on **scattering**. The physics behind scattering has its ground in the interaction between quantum objects and matter. The most prominent quantum objects in our focus are **neutrons**, **x-rays**, and **light**; they all obey the particle–wave dualism, which is connected by the De Broglie wavelength

$$\lambda = \frac{h}{mv} \tag{6.1}$$

If a quantum object hits a particle or **scattering center** in a sample, it is re-irradiated into space, and the overlay of multiple such scattered quantum objects that are re-irradiated by multiple adjacent scattering centers in a sample causes a **scattering intensity pattern**. Analysis of this pattern allows us to draw conclusions on the spatial distribution of the scattering centers in the sample, and with that, about the *structure* of the sample that we investigate. Depending on the wavelength of the scattered quantum objects, we can do so on different length scales. The three mentioned kinds of quantum objects can therefore all be used as "magnifying glasses" to provide structural information about the sample, with different extent of magnification depending on their wavelength. However, they do so from slightly different perspectives. This is

https://doi.org/10.1515/9783110672817-006

caused by the different ways in which **contrast** between the primary beam and the sample is generated during the scattering experiment. When light is used, the scattering contrast stems from spatial variation of the refractive index in the sample, which is correlated to spatial variation of the polarizability. For x-rays, contrast is caused by spatial distribution of electron densities, while for neutrons, contrast is caused from differences in the so-called scattering lengths of certain atoms in the sample, which is most pronounced for the isotope pair of hydrogen and deuterium. All these different sources of scattering eventually trace down to the spatial distribution of matter in the sample, or in other words, to the sample **nanostructure**, which can therefore be explored by scattering. To properly account for the different origins of contrasting in light, neutron, and x-ray scattering, special sample preparation is necessary in some of these methods. In the case of neutron scattering, we need to tag interesting parts of the samples by exchange of hydrogen atoms by deuterium atoms.[65] For x-ray scattering, we must have atoms with a high electron density in the sample. While this might seem tedious at first glance, it offers the chance to selectively **label** certain areas of a polymer, for example, its side chains or chain ends, and then to selectively investigate them. Often, combinations of the three scattering methods are used to generate a complimentary dataset.

A simple scattering experiment is the **Bragg diffraction** on a crystal lattice; it is schematically shown in Figure 97. An incoming beam causes each scattering

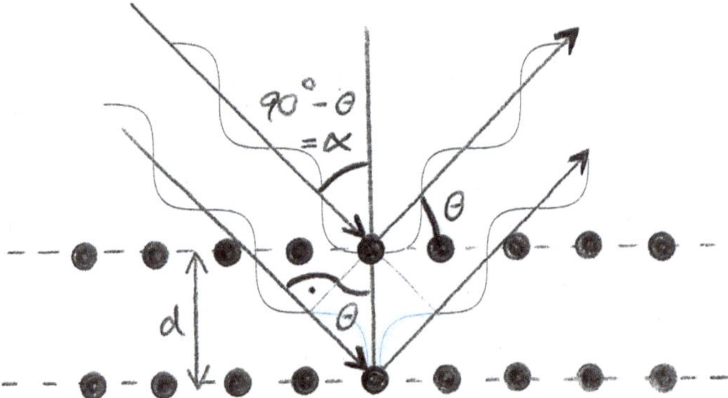

Figure 97: *Bragg* diffraction of a neutron or x-ray beam on crystal layers with a lattice constant of *d*. The part of the wave front scattered at the lower lattice plane has to travel further to reach the detector than the part scattered at the upper lattice plane. The extra distance traveled is highlighted in blue color. Following the geometry, it is $2d \sin \theta$. Note that due to the reflection at an interface of an optically thin to an optically thick medium, the phase of the wave is shifted by 180° at the reflection point.

65 The most simple way to realize that is to exchange the normal solvent by a deuterated one, as also used in NMR spectroscopy.

center (here: atoms in the sample) to re-irradiate incident x-rays as an isotropic spherical wave. When looking from a specific scattering angle, θ, we detect an overlay of all the radiation that is deflected into this specific direction. The sketch in Figure 97 shows that the part of the wave front scattered at the lower lattice plane has to travel further than the part scattered at the upper lattice plane. Following the geometry, the extra distance traveled is $2d \sin \theta$. This causes an interference that is positive and creates scattering peaks at the detector if that extra distance matches an integer multiple of the wavelength. This criterion is expressed in the **Bragg equation**

$$2d \sin \theta = n\lambda, \quad \text{with } n = 1, 2, 3, \ldots \tag{6.2}$$

An alternative way to sketch this scattering experiment is with the concept of a **wave vector**, as shown in Figure 98. Here, $\vec{k}_e$ is the wave vector of the entering beam and $\vec{k}_s$ the wave vector of the scattered beam. The difference of both vectors is called the **scattering vector** $\vec{q}$. The absolute values of both vectors are equal to

$$\left|\vec{k}_e\right| = \left|\vec{k}_s\right| = \frac{2\pi}{\lambda} \tag{6.3}$$

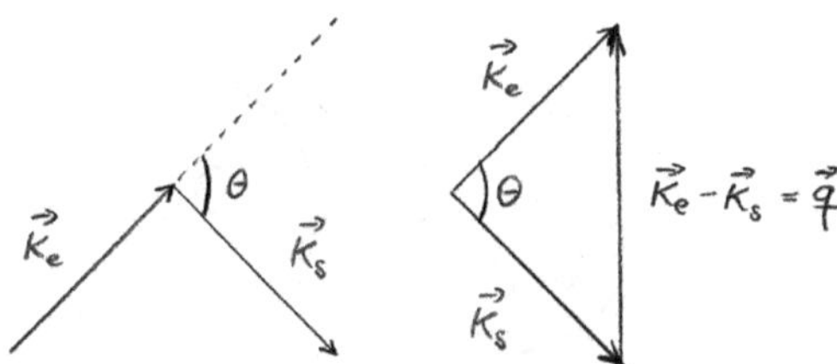

Figure 98: Vector representation of a scattering process. The entering wave vector is marked $\vec{k}_e$, whereas $\vec{k}_s$ denotes the scattered vector. The difference of both generates the *scattering vector* $\vec{q}$. Picture redrawn from J. S. Higgins, H. C. Benoît: *Polymers and Neutron Scattering*, Clarendon Press, Oxford, **1994**.

From geometry (see Figure 98), we obtain the scattering vector $\vec{q}$ as follows:

$$|\vec{q}| = q = \frac{4\pi}{\lambda} \sin(\theta/2) \tag{6.4}$$

With that, the Bragg condition for positive interference reads

$$\frac{2\pi}{q} = \frac{d}{n} \tag{6.5}$$

The equation shows how the structural feature d relates to the scattering vector q: the smaller the structures, the larger must q be to probe them. In other words, a low q-value means a low magnification, and vice versa.

6.2 Scattering regimes

The scattering vector, our "magnifying glass", can be varied by the wavelength of the scattered quantum objects. Neutrons and x-rays have short wavelengths (λ = 0.01–10 nm) and thereby realize high scattering vectors, whereas light has long wavelengths (green laser λ = 543 nm, red laser λ = 633 nm), and its scattering therefore covers small q-values. Further fine adjustments can be made by the scattering angle θ. This allows the investigation of various structural characteristics of polymeric systems, ranging from the entire polymer coil down to its local chemical structure. The accessible structural motifs of a polymer system are schematically shown in Figure 99. Light scattering probes the first two domains. In the first, our zoom level is so low that the polymer coils appear as pointlike masses, which allows for determination of their *molar mass* ("zero-angle domain"). In the second regime, zoomed in a little more, the polymer coil *dimensions* become visible (Guinier domain). To reach even more magnification, we have to switch from light to smaller wavelength quantum objects. By using either x-rays or neutrons, we zoom in a lot further and reach the third regime, in which parts of the chain become visible (if we observe at small scattering angles). Here, no more information on the chain size is gathered, and hence, the degree of

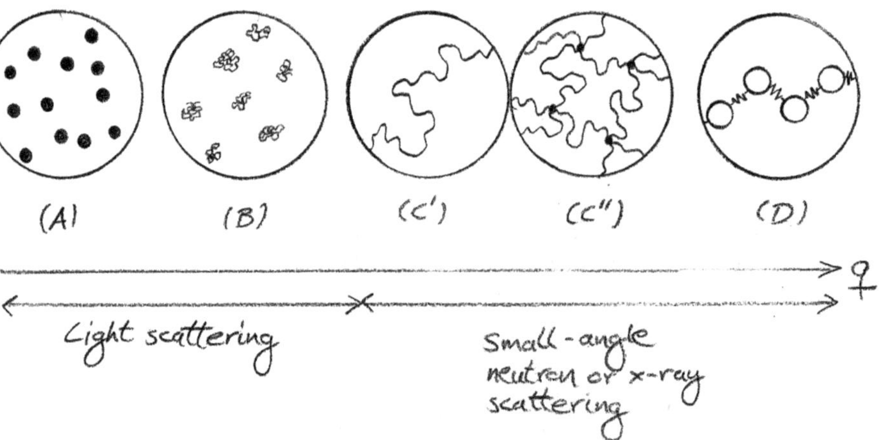

Figure 99: Scheme of the accessible scattering regimes. Light scattering can reach the first two domains, in which the polymer chains appear either pointlike (A) or, in the Guinier domain, as objects whose dimensions, shape, and mass can be determined (provided their radius of gyration is larger than about 50 nm) (B). Small-angle neutron or x-ray scattering allows for further magnification. Here, we lose information about degree of polymerization and polydispersity, but we gain knowledge about substructural characteristics such as the persistence length in the case of linear chains in a dilute solution (C') or the correlation length in the case of cross-linked or overlapping chains in a network at semidilute concentration (C"). At further magnification, we start to observe single chain segments only, until we finally resolve the local chemical structure of the polymer segmental units (D). Picture redrawn from J. S. Higgins, H. C. Benoît: *Polymers and Neutron Scattering*, Clarendon Press, Oxford, **1994**.

polymerization and the polydispersity of it becomes irrelevant. Instead, substructural characteristics such as the persistence length or, in solution, the solution-network correlation length are revealed. Even further magnification, as achieved at higher scattering angles, gives us a glimpse on single chain segments that may appear rodlike already if their persistence length is larger than the zoom level. Having reached the final regime, at even higher scattering angles, we can resolve the local chemical structure and gain information about chain tacticity and other (side-)chain orientations.

6.3 Structure and form factor

If we take Figures 97 and 99 together, we can guess already that there must be a general connection between the q-dependence of the scattering intensity, $I(q)$, and the structure of the sample. This is indeed the case, and that connection is given by the so-called **structure factor,** $S(q)$. It is defined as

$$I(q) = \Delta b^2 \; S(q) \tag{6.6}$$

The factor Δb is a measure of the difference in contrast to the species of interest and its surroundings. For light scattering, this is connected to relative polarizability differences. In small-angle x-ray scattering (SAXS), this is connected to relative electron-density differences that are especially strong if parts of the specimen are labeled with heavy atoms. For small-angle neutron scattering (SANS), contrast is generated by the so-called scattering-length density difference that is particularly pronounced between deuterium and hydrogen atoms. Hydrogen substitution for deuterium therefore allows for custom contrasting in SANS, but yet does not alter the polymer too much in its other properties.

The reduced form of the above equation describes the **differential scattering cross section** per unit volume:

$$\frac{1}{V} \cdot \frac{\partial \sigma}{\partial \Omega} = \frac{1}{V} \cdot \frac{I(\vec{q}) r^2}{I_0} \tag{6.7}$$

where σ denotes the scattering cross section, or number of scattered quantum objects per detector area, and Ω the solid angle under which the sample is observed. The expression on the right side of the equation, $\frac{1}{V} \cdot \frac{(I(\vec{q})r^2)}{I_0}$, is also called the **Rayleigh ratio.**

The structure factor, $S(q)$, can itself be split into two contributors. For neutron scattering, we have

$$S(q) = Nz^2 \; P(q) + N^2 z^2 \; Q(q) \tag{6.8a}$$

where N corresponds to the number of molecules of the species of interest in the scattering volume (volume cross section of beam and detector view), where each molecule contains z scattering centers. $P(q)$ is the **form factor** that accounts for *intra*molecular interferences happening when scattering waves originate from different scattering centers of the same (macro)molecule. It is directly connected to the shape of that molecule, and is therefore called form factor. $Q(q)$ accounts for *inter*molecular interferences.

In light scattering, by contrast, we have

$$S(q) = Nz^2\, P(q) \cdot Q'(q) \tag{6.8b}$$

where $Q'(q)$ accounts for interferences of light scattered from the centers of mass of different molecules. In the limit of a dilute solution, $Q(q) \to 0$ and $Q'(q) \to 1$, which simplifies both expressions to

$$S(q) = Nz^2\, P(q) \tag{6.8c}$$

The crux now lies in determining an analytical expression for $P(q)$ that matches the recorded dataset from the experiment. Note that at a concentration of $c \to 0$, when no intermolecular interference is possible, $P(q) \to 1$ for $q \to 0$. Hence, in the **zero-angle limit**, the scattering intensity becomes independent of the scattering vector q and reflects the number of molecules in the scattering volume as well as the number of scattering centers per molecule. This can then be translated to a molar mass, and this quantity can therefore be determined from the scattering intensity in the low-q limit. The mathematical definition of the form factor, $P(q)$, is given by

$$P(q) = \frac{1}{z^2} \sum_{i=1}^{z} \sum_{j=1}^{z} \langle \exp(-i\vec{q}\vec{r}_{ij}) \rangle \tag{6.9}$$

where $\vec{r}_{ij}$ is the vector that joins two scattering centers i and j in the molecule. The operator $\langle \ldots \rangle$ indicates averaging over all orientations and conformations. The orientational average of a function $f(\theta,\phi)$ is

$$\langle f(\theta,\phi) \rangle = \frac{1}{4\pi} \int_{\phi=0}^{2\pi} \int_{\theta=0}^{\pi} f(\theta,\phi) \sin\theta \mathrm{d}\phi \mathrm{d}\theta \tag{6.10}$$

When we apply this to eq. (6.9), we generate

$$P(q) = \frac{1}{z^2} \sum_{i=1}^{z} \sum_{j=1}^{z} \left\langle \frac{\sin qr_{ij}}{qr_{ij}} \right\rangle \tag{6.11}$$

For continuous bodies, the following alternative expression is also valid:

$$P(q) = \frac{1}{\int n(r)\,\mathrm{d}r} \int n(r) \left(\frac{\sin qr_{ij}}{qr_{ij}} \right) \mathrm{d}r \tag{6.12}$$

where $n(r)$ relates to the radial density function of the object of interest.

The simplest geometry that a scatterer can have is that of a regular sphere. Here, the scattering intensity does not depend on its orientation due to symmetry. This yields

$$P_{\text{sphere}}(q) = \left(\frac{3}{R^3}\int_0^R \frac{\sin qr}{qr} r^2 \mathrm{d}r\right)^2 = \frac{9}{(qR)^6}(\sin(qR) - (qR)\cos(qR))^2 \tag{6.13}$$

In that latter equation, the scattering vector appears as a dimensionless reduced variable qR, which means that the scattering vector q is normalized to the length scale of observation R. A typical result for a spherical object is shown in Figure 100. Note that there is no dependence on the scattering vector q at the very low-q end of the **Guinier region**. The scattering intensity there depends on the mass of spheres in the scattering volume, thereby allowing their molar mass to be estimated.

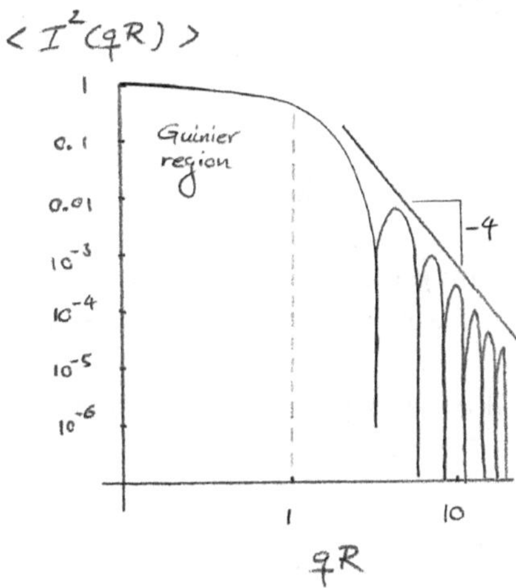

Figure 100: Sketch of the mean-square-average scattering intensity, $\langle|I(qR)|^2\rangle$, as a function of the reduced scattering vector, qR for spherical particles. In the low Guinier region ($qR \ll 1$), the scattering intensity does not depend on the scattering vector q, but only on the number of spheres in the scattering volume. In the region at higher q, by contrast, characteristic peaks appear that allow the size of the spherical particles to be determined.

A spherical object is easy to treat theoretically, because it is fully symmetric. For more complicated objects such as polymer coils, we have to simplify the derivation by looking onto the low-q and high-q extremes separately.

To capture the realities at low-q values corresponding to light scattering measurements, we can expand the basic formula given by eq. (6.11) into a Taylor series utilizing the expansion

$$\frac{\sin x}{x} = 1 - \frac{x^2}{3!} + \frac{x^4}{5!} - \cdots \tag{6.14}$$

This yields

$$P(q) = 1 - \frac{q^2}{3!z^2}\sum_{i=1}^{z}\sum_{j=1}^{z}\langle r_{ij}{}^2\rangle + \frac{q^2}{5!z^2}\sum_{i=1}^{z}\sum_{j=1}^{z}\langle r_{ij}{}^4\rangle - \cdots \tag{6.15}$$

In this series expansion, the second term gives the radius of gyration, R_g, which is defined as

$$R_g{}^2 = \frac{1}{z}\sum_{i=1}^{z}\langle r_i{}^2\rangle \tag{6.16a}$$

where r_i is the vector joining the scattering point i to the center of mass of the object.

An alternative expression in an ij two-point notation is

$$R_g{}^2 = \frac{1}{2z^2}\sum_{i=1}^{z}\sum_{j=1}^{z}\langle r_{ij}{}^2\rangle \tag{6.16b}$$

With that, we generate

$$P(q) = 1 - \frac{q^2}{3}R_g{}^2 + \cdots \qquad \text{(Zimm)} \tag{6.17}$$

The above equation allows us to construct a **Zimm plot**, which is the conventional means of deriving the radius of gyration R_g, the weight-average molar mass M_w, and the second virial coefficient A_2.

The form factor may also be approximated by an exponential to generate the so-called **Guinier function**

$$p(q) \cong \exp\left(-\frac{R_g{}^2q^2}{3}\right) \text{ for } qR_g < 1 \qquad \text{(Guinier)} \tag{6.18}$$

At high q-values, that is, in the regime of SANS and SAXS experiments, we resort to a scaling discussion. We retain the normalized scattering vector, qR, as a variable and use the following identity:

$$S(q) = Nz^2\, P(qR) = V\phi z\, P(qR) \tag{6.19}$$

where V is the sample volume, N the number of scattering particles therein, and ϕ the number of scattering units of degree of polymerization z per unit volume. Because we are only looking at a single, or possibly even just part of a single polymer chain, $N = 1$. Nz now directly corresponds to the number of scattering centers per chain. We can detect every Kuhn segment as a scattering center, and thus z corresponds to the degree of polymerization.

We assume that at large scattering vectors, $P(qR)$ has an asymptotic tail of type $(qR)^{-\alpha}$. Furthermore, we know several relations between the mass and size of polymers. For a Gaussian coil, it is given as $R \sim z^{1/2}$; for a chain in a good solvent, it is $R \sim z^{3/5}$; and for a rodlike chain, it is $R \sim z^1$. In general, we can say that the mass scales with the Flory exponent, $R \sim z^{\nu}$. Replacing that for R in the above equation yields

$$S(q) = V\phi z\, P(qz^{\nu}) = V\phi(qz^{\nu})^{-\alpha} = V\phi q^{-\alpha} z^{(1-\nu\alpha)} \tag{6.20}$$

At high q-values, the scattering intensity becomes independent of the degree of polymerization z, as we look into each coil but do not see it in its totality. This, however, is only fulfilled if $\alpha = \frac{1}{\nu}$, because only then will the z-factor be z^0. This yields q-dependencies for the aforementioned polymer geometries: $S(q) \sim q^{-2}$ for Gaussian chains, $S(q) \sim q^{-5/3}$ for polymer coils in a good solvent, and $S(q) \sim q^{-1}$ for maximally elongated polymeric rods. Generally, we can say that $S(q) \sim q^{-1/\nu}$.

Let us again come back to the most simple example of a regular sphere. For that geometry, in the higher q regime,[66] scattering can only come from the sphere's surface, because if we look completely inside or outside of it, there will be no scattering as there is no spatial contrast difference. As a result, the scattering intensity is proportional to the extent of the surface in the scattering volume. For the surface S we can express $R \sim S^{0.5}$ and $Nz^2 = NV^2 = NS^3$. Plugging this into eq. (6.8b), again under assumption of an asymptotic power-law tail, yields

$$S(q) = Nz^2\, P(qR) = NS^3 \left(qS^{0.5}\right)^{-\alpha} \tag{6.21}$$

Thus, the scattering intensity must be proportional to the total surface of the sample in the scattering volume, which is proportional to NS. That can only be fulfilled at $\alpha = 4$.

$$S(q) \sim q^{-4} \tag{6.22}$$

This is called **Porod's law** and visualized in Figure 100 by the envelope slope of −4.

66 To be precise: for $q > 1/R$, denoted as the Mie regime.

6.4 Light scattering

Lesson 18: Light Scattering on Polymers

The structure of many polymer systems can be assessed by different scattering methods, each with its specific pros and cons. The most widespread technique is light scattering. The following lesson sheds a special view to this method and shows how light scattering can serve to assess a polymer's molar mass, shape, and interactions with the surrounding.

Based on the above general thoughts about scattering, we now look on how this is reflected in actual experiments. We focus on light scattering, as this method is much easier to realize than neutron scattering or x-ray scattering, and as it is less demanding in view of the sample composition, in a sense that no heavy element or isotope labeling is necessary. According to the discussion of the scattering vector q in the last lesson, light scattering covers the zero-angle and Guinier regimes. This means that information about the molar mass and the size of the polymer coils is accessible.

6.4.1 Static light scattering

As we have learned earlier, light scattering generates its contrast from the relative differences of the refractive indices of the sample and its environment. But how does the light scattering process itself work? It starts with an incident, polarized light beam that hits a polarizable object, that is, a scattering center. There, the light induces a dipole moment μ according to

$$\mu_{\text{ind}} = \alpha \cdot E_{\text{incident}} = \alpha \cdot E_0 \cos(\omega t) \tag{6.23}$$

where α is the polarizability of the object under investigation. Due to that induced dipole moment, the scatterer starts to vibrate, and as a result, it emits radiation into all directions. The scattered electrical field can be estimated according to

$$E_s \sim \frac{1}{r} \sin\varphi \cdot \frac{\partial^2 \mu_{\text{ind}}}{\partial t^2} \sim \frac{1}{r} \sin\varphi \cdot E_0 \alpha \omega^2 \cdot - \cos(\omega t) \tag{6.24}$$

where $\left(\frac{1}{r}\right) \sin\varphi \cdot E_0 \alpha \omega^2$ is the amplitude of the scattered light that depends on the distance, angle, and oscillation frequency of the source; it corresponds to $E_{0,s}$. The expression $-\cos(\omega t)$ denotes the periodicity of the scattered light that is also the periodicity of the incident light. The intensity of the scattered light scales with the square of the electrical field, $I \sim E^2$. This yields

$$\left(\frac{E_{0,s}}{E_0}\right)^2 = \frac{I_s}{I_0} \sim \alpha^2 \omega^4 \tag{6.25}$$

We realize that the scattering intensity has a strong dependence on the frequency. Blue light is scattered much stronger than other colors. This is the reason why we see that part of the sunlight majorly when looking at the sky, which therefore appears blue to us[67] (while the sun appears yellow, because we see the remaining spectrum without the blue when looking at it).[68] The exact expression is given by

$$\frac{I_s}{I_0} = \frac{\pi^2 \cdot \alpha^2 \cdot \sin^2 \varphi}{\varepsilon_0{}^2 \cdot r^2 \cdot \lambda^4} \tag{6.26}$$

where ε_0 is the vacuum dielectric constant and λ the wavelength of the utilized light.

Now that we know how to treat a single particle, we can expand our view to multiple-particle systems. The easiest such system is that of a dilute ideal gas, because it contains N particles of the same nature as discussed above (spheres). Consequently, we only have to extend the last equation by the number of particles, N:

$$\frac{I_{s,\,\text{total}}}{I_0} = N \cdot \frac{\pi^2 \cdot \alpha^2 \cdot \sin^2 \varphi}{\varepsilon_0{}^2 \cdot r^2 \cdot \lambda^4} \tag{6.27}$$

The opposite to a dilute ideal gas is an ideal crystal. Such a material, however, cannot be investigated by light scattering. This is because the distance of the lattice planes is in the range of some picometers, whereas the wavelength λ of the utilized visible light is a couple of hundreds of nanometers. This causes each beam scattered in any direction to *always* find a counterpart point source that causes destructive interference, and therefore extinction of the scattered light. This is the reason why crystals are transparent. They might at most be colored, but this is caused by light absorption rather than scattering.

At a first glance, the latter conceptual picture is also similar for liquids: although there is no perfect order like in a crystal, the molecules in a liquid are still so densely packed that their average distances from one another are still much smaller than the light's wavelength. Again, as a result, the light is extinguished and the liquid has a transparent appearance. In contrast to a crystal, however, density *fluctuations* can occur due to the Brownian motion of the molecules. As a result, not *all* of the scattered rays interfere with an extincting counterpart at a given time, and

67 The oceans appear blue, too, because they reflect the blue color of the sky above them.

68 To be precise, the sky actually appears blue to us only when the light passes a short distance and is therefore scattered not too often, which is the case at the height of the day. In the morning and evening, by contrast, the light path is longer, so the blue components are scattered away completely, and we see the remaining red-orange parts at dawn and dusk.

so we do observe some scattering. As an example, consider liquid water: it scatters 200 times stronger than the gaseous water. It should be kept in mind, however, that the particle-number density in liquid water is 1200 times higher than in gaseous water. The "normalized" scattering therefore is only 1/6 times higher, so there is in fact still notable extinction due to destructive interference.

The polarizability of a molecule is usually anisotropic. If the molecule rotates and vibrates, this manifests itself such as if the polarizability α would oscillate:

$$\alpha = \alpha_0 + \alpha_k \cos(\omega_k t) \tag{6.28}$$

where ω_k is the eigenfrequency of the molecule that is closely connected to its rotational and vibrational eigenfrequencies. We can insert the above expression into eq. (6.23) to generate

$$\begin{aligned} \mu_{\text{ind}} &= \alpha \cdot E_{\text{incident}} = \alpha_0 \cdot E_{\text{incident}} + \alpha_k \cdot \cos(\omega_k t) \cdot E_{\text{incident}} \qquad | E_{\text{incident}} = E_0 \cdot \cos(\omega t) \\ &= \alpha_0 \cdot E_0 \cdot \cos(\omega t) + \alpha_k \cdot E_0 \cdot \cos(\omega t) \cdot \cos(\omega_k t) \\ &= \alpha_0 \cdot E_0 \cdot \cos(\omega t) + \frac{1}{2}\alpha_k \cdot E_0 \cdot [\cos(\omega - \omega_k)t + \cos(\omega + \omega_k)t] \end{aligned} \tag{6.29}$$

where the first term $\alpha_0 \cdot E_0 \cdot \cos(\omega t)$ accounts for the **Rayleigh scattering** that is a predominantly **elastic scattering** by particles much smaller than the wavelength of the radiation. Elastic scattering means that the frequency of the scattered light is the same as that of the incident light, which means that Rayleigh scattering is not accompanied by any energy exchange between the photon and the scatterer. The other two terms account for **Raman scattering**, which encompasses the fraction of the light that is scattered *inelastically*, whereby the scattered photons have a frequency and energy different from those of the incident photons. When the scatterer absorbs energy and the scattered photon has less energy than the incident one, we have ***Stokes* Raman scattering**, whereas an energy transfer to the scattered photon so that its energy is higher than that of the incident one is called ***anti-Stokes* Raman scattering**. More important for static light scattering of liquids or solutions is the Rayleigh scattering part of the process, which is why we limit our further discussion to this type.

Let us first examine the case of a dilute polymer solution. Its refractive index, n_{solution}, is given by

$$n_{\text{solution}} = n_{\text{solvent}} + \frac{dn}{dc}c \overset{\text{square}}{\Rightarrow} n^2_{\text{solution}} = n^2_{\text{solvent}} + 2n_{\text{solvent}}\frac{dn}{dc}c + \left(\frac{dn}{dc}c\right)^2 \tag{6.30}$$

Due to its negligible influence, we will disregard the final term of this expression, $\left(\frac{dn}{dc}c\right)^2$.

The refractive indices n are connected to the polarizability α via the following expression:

$$n^2 - 1 = \frac{(N/V)}{\varepsilon}\alpha \tag{6.31}$$

Applied to the refractive index of the solution, n_{solution}, we generate

$$n^2_{\text{solution}} - 1 = \frac{\left(\frac{N_{\text{solvent}}}{V}\right) - \left(\frac{N_{\text{solute}}}{V}\right)}{\varepsilon_0}\alpha_{\text{solvent}} + \frac{\left(\frac{N_{\text{solute}}}{V}\right)}{\varepsilon_0}\alpha_{\text{solute}} \tag{6.32}$$

The corresponding expression for the refractive index of the solvent, n_{solvent}, is

$$n^2_{\text{solvent}} - 1 = \frac{\left(\frac{N_{\text{solvent}}}{V}\right)}{\varepsilon_0}\alpha_{\text{solvent}} \tag{6.33}$$

By subtracting the former two expressions from one another and taking eq. (6.30) into account, we can derive a formula that connects the refractive index differences to the incremental refractive index change with concentration, $\frac{\mathrm{d}n}{\mathrm{d}c}$:

$$n^2_{\text{solution}} - n^2_{\text{solvent}} = \frac{\left(\frac{N_{\text{solute}}}{V}\right)}{\varepsilon_0}(\alpha_{\text{solute}} - \alpha_{\text{solvent}}) = 2n_{\text{solvent}} \cdot \frac{\mathrm{d}n}{\mathrm{d}c}c \tag{6.34}$$

Solving this for an equation describing the polarizability difference, $\Delta\alpha = \alpha_{\text{solute}} - \alpha_{\text{solvent}}$, yields

$$\Delta\alpha = 2n_{\text{solvent}} \cdot \varepsilon_0 \frac{\mathrm{d}n}{\mathrm{d}c}\frac{c}{N/V} = 2n_{\text{solvent}} \cdot \varepsilon_0 \frac{\mathrm{d}n}{\mathrm{d}c}\frac{M}{N_\mathrm{A}}, \quad \text{with } \frac{N}{V} = \frac{c \cdot N_\mathrm{A}}{M} \tag{6.35}$$

We can now insert this expression into the Rayleigh formula from above (eq. 6.27):[69]

$$\begin{aligned}
\frac{I_{\text{s, total}}}{I_0} &= N \cdot \frac{\pi^2 \cdot 4n^2_{\text{solvent}} \cdot {\varepsilon_0}^2 \left(\frac{\mathrm{d}n}{\mathrm{d}c}\right)^2 M^2}{{N_\mathrm{A}}^2 \cdot {\varepsilon_0}^2 \cdot r^2 \cdot \lambda^4} \cdot \sin^2\varphi \quad \Big| N = \frac{cN_\mathrm{A}V}{M} \\
&= \frac{cN_\mathrm{A}V}{M} \cdot \frac{\pi^2 \cdot 4n^2_{\text{solvent}} \cdot {\varepsilon_0}^2 \left(\frac{\mathrm{d}n}{\mathrm{d}c}\right)^2 M^2}{{N_\mathrm{A}}^2 \cdot {\varepsilon_0}^2 \cdot r^2 \cdot \lambda^4} \cdot \sin^2\varphi \\
&= c \cdot M \cdot V \cdot \left(\frac{\mathrm{d}n}{\mathrm{d}c}\right)^2 \cdot \frac{4\pi^2 \cdot n^2_{\text{solvent}} \cdot \sin^2\varphi}{N_A \cdot r^2 \cdot \lambda^4}
\end{aligned} \tag{6.36}$$

69 Equation (6.27) has α in it, because it holds for gases, where refractive index differences occur because we either have a gas molecule (with polarizability α) or nothing in a spot; in the case of solutions, we need to work with $\Delta\alpha$ instead, because then, refractive index differences occur because we either have a solvent molecule (with polarizability α_{solvent}) or a solute molecule (with polarizability α_{solute}) in a spot.

While this equation might seem complicated at first glance, all parameters except for the concentration c and the molar mass M are constant for a given system and setup. This allows us to determine the molar mass M of a compound by performing concentration-dependent static light scattering measurements.

As light is scattered into all directions, there is a trivial sample–detector distance dependence of r^{-2}. Furthermore, the scattering volume V has a trivial influence on the scattering intensity detected: the larger the volume, the larger is the relative intensity, $\frac{I_{\mathrm{s,\,total}}}{I_0}$. To get experimental setup-independent results, we normalize to these two factors and introduce the **Rayleigh ratio**, R_θ, that has the unit m^{-1}:

$$R_\theta = \frac{I_\theta r^2}{I_0 V} = \frac{4\pi^2 n^2}{N_\mathrm{A}\lambda^4}\left(\frac{\mathrm{d}n}{\mathrm{d}c}\right)^2 \sin^2\varphi \cdot cM \tag{6.37a}$$

We can simplify this expression greatly by combining all the constant values into a constant K. This yields

$$R_\theta = \frac{I_\theta r^2}{I_0 V} = K \cdot c \cdot M \tag{6.37b}$$

In a polydisperse sample, all components with molar mass M_i and the corresponding concentration c_i scatter independently, yielding:

$$R_\theta = K \sum_i c_i M_i \tag{6.38}$$

We know from Section 1.3.2 that the weight-average molar mass of a polymer is defined as

$$M_\mathrm{w} = \frac{\sum_i N_i M_i^2}{\sum_i N_i M_i} = \frac{\sum_i c_i M_i}{\sum_i c_i} \tag{6.39}$$

Inserting this into the expression for the Rayleigh ratio, R_θ, yields

$$\frac{Kc}{R_\theta} = \frac{1}{M_\mathrm{w}} \tag{6.40}$$

With this equation, we can directly calculate the weight-average molar mass from the Rayleigh ratio measured during a scattering experiment.

An alternative viewpoint can be gained from fluctuation theory, which combines thermodynamics with light scattering. To account for and quantify concentration fluctuations, which are the actual cause of scattering from solutions, we regard our system to be composed of many elementary cells δV, that are large compared to molecular scales but smaller than $\left(\frac{\lambda}{20}\right)^3$. Concentration fluctuations occur due to molecular exchange between these cells, and are described by

$$c = \bar{c} + \delta c \;\Rightarrow\; \alpha = \bar{\alpha} + \delta\alpha \tag{6.41}$$

Note in this context that strictly speaking, there are fluctuations of the polarizability in each elementary cell, α, not only due to concentration fluctuations, but also due to other types of fluctuations in the system, such as temperature and pressure:

$$\delta\alpha = \left(\frac{\partial\alpha}{\partial c}\right)_{p,T} \partial c + \left(\frac{\partial\alpha}{\partial p}\right)_{c,T} \partial p + \left(\frac{\partial\alpha}{\partial T}\right)_{c,p} \partial T \tag{6.42}$$

However, these fluctuations affect both the solvent and the solute equally, so we may disregard them.

Going back to the original scattering equation (6.27), we need an expression for α^2. It is calculated from eq. (6.42) as

$$\alpha^2 = (\bar{\alpha} + \delta\alpha)^2 = \bar{\alpha}^2 + 2\bar{\alpha}\delta\alpha + (\delta\alpha)^2 \tag{6.43}$$

Let us examine the three terms of this expression a little closer. The first one, $\bar{\alpha}^2$, is the same for all volume elements very much like in an ideal crystal, so there is no contribution due to interference. The second term, $2\bar{\alpha}\delta\alpha$, can either have a positive or a negative value, but it will be zero upon averaging. This only leaves the third term $(\delta\alpha)^2$ to actually cause the observed scattering intensity. By inserting this last term into the Rayleigh equation, we generate

$$\frac{I_{\mathrm{s,total}}}{I_0} = N \cdot \frac{\pi^2 \cdot (\delta\alpha)^2 \cdot \sin^2\varphi}{\varepsilon_0{}^2 \cdot r^2 \cdot \lambda^4} \tag{6.44}$$

Note that the $\sin^2\varphi$ term is due to light polarization in this case.

The mean-square fluctuation of the polarizability α is connected to the refractive indices n as follows:

$$\langle\delta\alpha^2\rangle \sim n^2 \left(\frac{\mathrm{d}n}{\mathrm{d}c}\right)^2 \langle\delta c^2\rangle \tag{6.45}$$

The mean-square fluctuation of the concentration, $\langle\delta c^2\rangle$, is given by statistical thermodynamics as

$$\langle\delta c^2\rangle = \frac{RTc}{\left(\frac{\partial\pi}{\partial c}\right)_T} \tag{6.46}$$

Plugging this into our above formula and using the virial series expansion of the concentration-dependent osmotic pressure, we get:

$$\frac{Kc}{R_\theta} = \frac{1}{M_w} + 2A_2c + \cdots \tag{6.47}$$

where A_2 is the second virial coefficient of the osmotic pressure, obtained as a z-average.

For molecules that are larger than a twentieth of the used wavelength, $\frac{\lambda}{20}$, or roughly 20 nm, there will be additional *intra*molecular interference due to optical path differences as sketched in Figure 101. This leads to an additional angular dependence of the scattered light, and necessitates to account for an angular-dependent particle form factor, $P(\theta)$:

$$\frac{Kc}{R_\theta} = \frac{1}{M_w P(\theta)} + 2A_2 c + \cdots \tag{6.48}$$

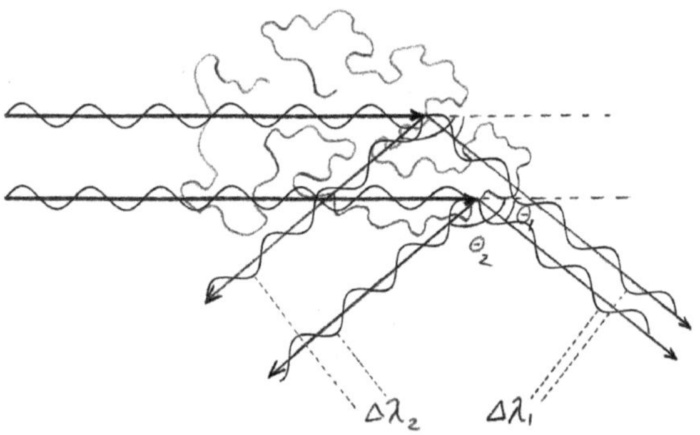

Figure 101: Schematic of the optical path difference, $\Delta\lambda$, of two interfering beams scattered at different points of a polymer coil with different angles of observation. The lower left pair of beams has a bigger scattering angle, resulting in a bigger optical path difference, whereas the lower right pair of beams has a smaller scattering angle and therefore a smaller optical path difference.

We can use the formula for the form factor that we derived in the last section (eq. 6.17) and insert into it the expression for the scattering vector q (eq. 6.4) to generate

$$P(\theta) = 1 - \frac{1}{3} q^2 \langle R_g{}^2 \rangle = 1 - \frac{16\pi^2 n^2}{3\lambda^2} \langle R_g{}^2 \rangle \sin^2 \frac{\theta}{2} \tag{6.49}$$

Note that $\sin^2\theta$ here is *not* due to light polarization, but comes from the definition of the scattering vector q.

Inserting the above expression into eq. (6.48) yields

$$\frac{Kc}{R_\theta} = \frac{1}{M_w} \left(1 + \frac{16\pi^2 n^2}{3\lambda^2} \langle R_g{}^2 \rangle \sin^2 \frac{\theta}{2} \right) + 2A_2 c \tag{6.50}$$

Measuring the Rayleigh ratio R_θ as a function of the scattering angle θ and the concentration c allows three parameters to be determined simultaneously: the weight-average molar mass M_w, the radius of gyration R_g, and the second virial coefficient

A_2. This determination is often done graphically using a **Zimm plot**. For this purpose, experiments are performed at several angles that all satisfy the condition $qR_g < 1$, and at least at four concentrations. Instead of creating two graphs for the angle- and the concentration-dependent series, both can be plotted in a single graph, as shown in Figure 102. From the linear extrapolation of the angle-dependent series as well as the concentration-dependent series to $\theta \to 0$ and $c \to 0$, respectively, the weight-average molar mass can be determined. The radius of gyration can be calculated from the slope of the angle series at the zero concentration, and the second virial coefficient can be calculated from the slope of the concentration series at the zero angle.[70]

In statistical thermodynamics, the mean-square concentration fluctuation may also be given as

$$\langle \delta c^2 \rangle = \frac{RTc}{\left(\frac{\partial^2 g}{\partial c^2}\right)_{p,T}} \quad \text{with } g = \frac{\partial G}{\partial V} \quad \text{(Gibbs free energy per volume element)} \tag{6.51}$$

From this expression, we see that

$$\frac{I_s}{I_0} \sim \frac{RTc}{\left(\frac{\partial^2 g}{\partial c^2}\right)_{p,T}} \tag{6.52}$$

At a certain critical demixing point, we have a horizontal slope of $G(c)$, meaning that $\left(\frac{\partial^2 G}{\partial c^2}\right) = 0$, as shown in Figure 103. At this point we observe very strong scattering, the so-called critical opalescence.

70 Note, in this context, that the quantities R_g and A_2 are obtained as z-averages from the Zimm plot, whereas the molar mass is obtained as a w-average, M_w. This is due to the following reasoning. In general, the electrical field amplitude, E, of a single light-scattering particle (or of a polymer coil in the limit of $\theta \to 0$, where intramolecular interferences are irrelevant) is proportional to its mass: $E \sim m$ (or, if you wish, to its molar mass: $E \sim M$). The intensity of the scattered light scales with the square of that electrical field amplitude: $I \sim E^2$. This means that $I \sim m^2$ (or $I \sim M^2$ if you like that better). For an ensemble of N particles, we get $I \sim N \cdot m^2$ (or, if we express it in molar numbers, $I \sim (n/N_A) \cdot M^2$, with N_A the Avogadro number), which already resembles the basic definition of a quantity's z-average (whereas that of a w-average is related to $n \cdot M^1$ and that of an n-average relates to $n \cdot M^0$). As a result, any quantity obtained from light scattering is generally received as a z-average, including the hydrodynamic radius R_H (that is determined from dynamic light scattering, as detailed in the following section), the radius of gyration R_g, and the second virial coefficient A_2. The molar mass, however, is determined from the Zimm equation, $\frac{Kc}{R_\theta} = \frac{1}{M_w}$ (6.40) (which reflects the Zimm plot in the limit of $\theta \to 0$ and $c \to 0$). The concentration in the numerator equals to $c = \frac{\sum_i n_i M_i}{V}$, whereas the Rayleigh ratio in the denominator equals to $R_\theta = K \sum_i c_i M_i$ according to eq. (6.38). Together, this gives $R_\theta = K \frac{\sum_i n_i M_i M_i}{V} = K \frac{\sum_i n_i M_i^2}{V}$. Plugging both into the Zimm formula gives $\frac{\sum_i n_i M_i}{\sum_i n_i M_i^2} = \frac{1}{M_w}$, which directly reflects the definition of a w-average for the molar mass according to eq. (6.39).

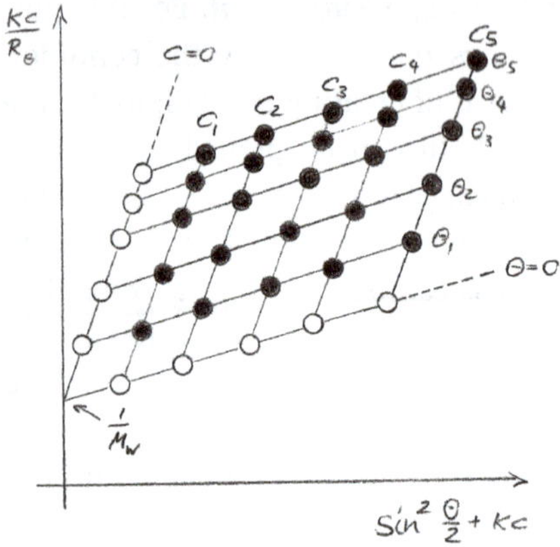

Figure 102: Plot of $\frac{Kc}{R_\theta}$ as a function of $\sin^2\left(\frac{\theta}{2}\right)+kc$, the so-called *Zimm plot*. The constant k can be chosen arbitrarily to make the plot look clear in a sense that the data points are spread equally in the x- and y-directions. From the common y-intercept of the extrapolations for $\theta \to 0$ and $c \to 0$, the weight-average molar mass, M_w, can be determined. On top of that, the slope of the $\theta \to 0$ extrapolated line gives the second virial coefficient, A_2, whereas the slope of the $c \to 0$ extrapolated line gives the radius of gyration, R_g. Picture redrawn from B. Tieke: *Makromolekulare Chemie*, Wiley VCH, **1997**.

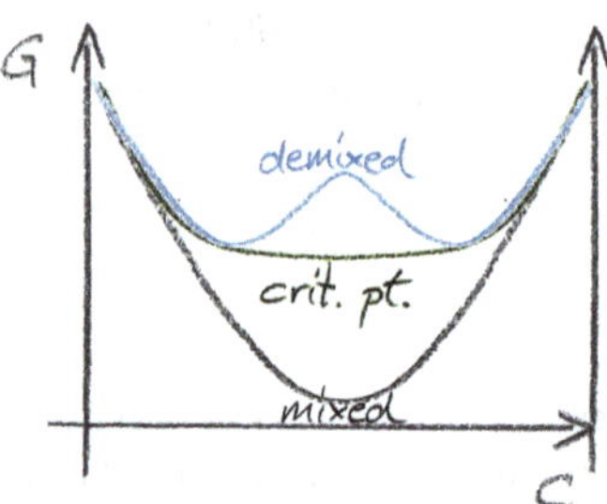

Figure 103: At the critical demixing point, the slope of $G(c)$ is horizontal, such that very intense scattering, called critical opalescence, occurs.

6.4.2 Dynamic light scattering

When light scattering measurements are carried out in a time-dependent fashion, we talk about **dynamic light scattering**. With this method, as its name implies, we can obtain information about the dynamics of a sample. This is because scattering centers in the probe volume are subject to Brownian motion: they constantly change their mutual distances, thereby causing temporal *fluctuation* of the scattering intensity due to temporal fluctuation of the *inter*molecular interference. A time-dependent graph of these fluctuations is shown in Figure 104(A). The fluctuating scattering intensity can be mathematically treated by **autocorrelation**:

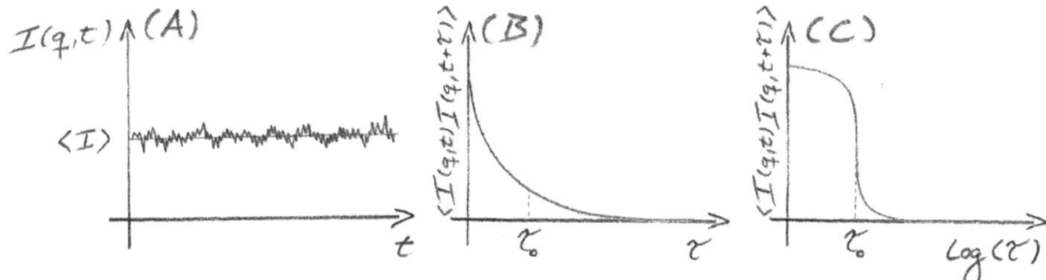

Figure 104: Data processing in dynamic light scattering. (A) Time-dependent intensity fluctuations as recorded by dynamic light scattering. (B) Intensity autocorrelation as a function of τ, shown both in a linear and in a (C) semilogarithmic plot. The characteristic time τ_0 corresponds to the decay of the autocorrelation function to $1/e$ of its original value and can be easily spotted in the semilogarithmic representation as the inflection point of the function.

$$g_2(\tau) = \frac{\langle I(t)I(t+\tau)\rangle}{\langle I(t)\rangle^2} \tag{6.53}$$

The index 2 indicates that this is an **intensity autocorrelation function.** An index of 1 denotes the electrical field autocorrelation function, with E instead of I. In such an autocorrelation analysis, we shift a duplicate of the fluctuation intensity signal along a conceptual time axis by a lag time τ. The further we shift, the less similar is the duplicated signal to the original. As a result, the autocorrelation function $g_2(\tau)$ decays, as shown in Figure 104(B). A characteristic point is the decay to $1/e$ of the original at the characteristic time τ_0, as also indicated in Figure 104(B). It is easily determined as the inflection point in a semilogarithmic plot, as shown in Figure 104(C). At this time, the scattering centers have displaced themselves by a distance of $1/q$. This results in the following Einstein–Smoluchowski analog equation:

$$\frac{1}{q^2} \approx D\tau_0 \Leftrightarrow \frac{1}{\tau_0} = \Gamma \approx Dq^2 \tag{6.54}$$

Measuring τ_0 at different scattering vectors q, that is, at different scattering angles, shows whether the motion in the sample is diffusive or not. For diffusive motion, we need to find a linear interdependence of $1/\tau_0$ and q^2. If this is the case, then we can calculate the prime resulting parameter obtained by DLS: the **diffusion coefficient** D. Using the Stokes–Einstein equation, we can translate this diffusion coefficient into a hydrodynamic radius, R_h. This is not, however, the precise hydrodynamic radius of the sample, but that of an imaginary, perfectly spherical particle that diffuses ideally and has the same diffusion coefficient as the one obtained from the experiment. For polymers, this assumption is valid most of the time, because the shape of a Gaussian coil can be assumed to be spherical on timescales longer than the Rouse or Zimm time.

7 Closing remarks

The target of this book is to deliver an **understanding of the relations between *structure* and *properties* of polymers**, thereby **bridging** the fields of **polymer chemistry**, which focuses on the primary molecular-scale structure of individual polymer chains, and **polymer engineering**, which focuses on the macroscopic properties of polymer materials. With that arc, **molecular parameters** are **rationally connected** to **macroscopic function**. As a fundament for this bridge, the first chapters of this book have considered the shape of a single ideal polymer coil using models such as the *phantom chain model* and the *Kuhn model*. From that, we have recognized that the statistics of the ideal chain are those of a *random walk*, and we have used this notion to quantify the *free energy* and the mechanical characteristics upon *deformation* of such ideal chains. We have then moved further to incorporate chain–chain and chain–solvent interactions into our model, thereby focusing on *real polymer chains*; this enabled us to categorize solvents based on their ability to dissolve and swell a polymer. We have realized that when all interactions exactly balance each other, the real chain acts like an ideal chain, and we have called this quasi-ideal state the *Θ-state*. In a next step, we have developed a thermodynamic theory for polymer solutions, the *Flory–Huggins theory*, based on a mean-field approach, and with it we were able to construct a full phase diagram of a polymer solution or a polymer blend. Furthermore, we have learned about two conceptual approaches to model the dynamics and motion of polymer chains: the Rouse and the Zimm model.

While doing all of the above, we have often encountered striking similarities of the concepts that we have employed to those known from elementary physical chemistry. These similarities appeared both on the conceptual and on the mathematical side of our argumentations. Conceptually, for example, the description of an ideal polymer chain and an ideal gas was based on the same assumptions. The same resemblance was also retained for real polymer chains and real gases, which are both subject to mutual and environmental interactions. As a consequence, both materials display quasi-ideal states at a very specific temperature, the Θ-temperature for polymer chains and the Boyle temperature for gases. Mathematically, we have seen similarities between the end-to-end distance distribution within a Gaussian polymer coil, the Boltzmann and the Maxwell–Boltzmann distributions of kinetics of gas particles, and the electron density distribution in an *s*-orbital. We have also seen how random-walk statistics that describes the diffusive motion of a particle or molecule also applies to the statistical treatment of the shape of ideal polymer chains.

Methodically, we have touched upon two fundamental approaches that are used in polymer physical chemistry. First, we have learned how a multibodied complex system can be simplified by only considering average values. This is the *mean-field* approach. Second, we are now familiar with the concept of scaling laws.

https://doi.org/10.1515/9783110672817-007

Realizing that polymers are self-similar objects, we can rescale our conception of them to different length scales without having to change the mathematical description of them, as it is done in the *blob concept*.

When we reconsider the conceptualized goal of this book, to build a bridge between *polymer chemistry* and *polymer engineering*, we may conclude that the first four chapters of this book have laid its pillars. Based upon that, the actual construction of the bridge has been outlined in the fifth chapter of this book. In this central chapter, we could rationally and quantitatively understand why many of the polymeric materials that we encounter in our everyday lives behave the way they do. Our prime focus in this consideration was on those properties of polymers that have made greatest impact on our lives in the past decades, which is their *mechanical* properties, spanning from viscous flow to elastic snap, along with their time and temperature dependence. To complete our picture, this book has then been closed with a sixth chapter on *scattering methods* as a prime experimental platform to characterize nano- and microstructures of polymer systems. With that, we have closed the circle of our endeavor of deriving relations between the structure of a polymer system and its properties (primarily the mechanical ones).

Index

https://doi.org/10.1515/9783110672817-008

www.ingramcontent.com/pod-product-compliance
Lightning Source LLC
Chambersburg PA
CBHW061412210326
41598CB00035B/6191

* 9 7 8 3 1 1 0 6 7 2 8 0 0 *